Michael Bowles was born in 1940 at the Bird in Hand pub at Stoulton near Worcester, and spent the next few months totally unaware of the Dunkirk evacuation, the Battle of Britain, the blitz, and the suicide of Virginia Woolf.

He received primary education at Bredicot village school, including unscheduled, unauthorised sex education which was not well received. This was followed by secondary education at Pershore Secondary Modern School. During both he was boringly serious and well-behaved.

Then came the misadventures of a student engineering apprenticeship at the Royal Radar Establishment and College of Electronics at Malvern where he was less well behaved. Next came a short period as a design draughtsman before obtaining a graduate engineer post at the Royal Aircraft Establishment at Farnborough leading to a career as a chartered engineer with MoD Procurement Executive and the Defence Evaluation and Research Agency.

He now lives in retirement in Blandford Forum engaged in his hobbies of writing and art. He still does not know if he is any good at either.

This is for Dora, Carole and Irene, the ladies in my life.

Michael Bowles

WORK AND PLAY

TALES OF AN UNREMARKABLE ENGINEER

AUSTIN MACAULEY PUBLISHERS™

LONDON • CAMBRIDGE • NEW YORK • SHARJAH

A CIP catalogue record for this title is available from the British Library.

ISBN 9781398477551 (Paperback)
ISBN 9781398477568 (ePub e-book)

www.austinmacauley.co.uk

First Published 2024
Austin Macauley Publishers Ltd®
1 Canada Square
Canary Wharf
London
E14 5AA

I cannot think of helpers in the research (there wasn't any research, just my memories) but some may crop up during your processes.

However, I should perhaps thank the people who helped me in my career and during crises in my life who are mentioned in the book. (i.e., Mr Howell, "Happy" Partridge, and George Smith.

Table of Contents

Introduction

I was sorting through my document files recently when I came across a copy of my last CV which was compiled in February 2000, when I was being considered for a consultant post at Boscombe Down after my retirement. The idea occurred to me to use this as the basis for a book by turning it upside down with the earliest entries first and using them as chapter headings.

I could then tell the stories of what really happened at the time. I also took the liberty of straying outside the strict limits implied by the CV to include such items as the Rag Weeks of my student days at Malvern, a holiday in Brussels for the World Fair of 1958, my hobbies and milestones in my life outside work like courtship, marriage, children and grandchildren.

For the various posts I held, I have usually used descriptive titles rather than the weird and wonderful civil service grades/ranks. I have also tried to keep technical descriptions in check and concentrate more on the character of colleagues and acquaintants encountered on the way.

I have inserted a relevant extract of the CV at the start of the chapter concerned. The final CV extract covering resource management is much abbreviated from the original as it is a bit long winded.

I have taken the liberty of purloining some bits of my earlier book 'Oh Carole' and inserting them where appropriate.

One may wonder why the aircraft pictures are 'artists' impressions' rather than photographs. In fact, I intended to use photographs, but my approach to publishers of the sources of these photographs for permission to use them received no response whatsoever. Also, I could not find a photograph of Canberra WH 953 in its mid-1960s 'TSR2 radar' configuration. Some of the pictures are 'composites' composed from two or three photographs.

These pictures were created during the Covid-19 lockdown. When the lockdown started, my wife, Irene started badgering me to go back to doing some art, but I was reluctant at first only giving way in the end just to shut her up.

However, once I started, I went into an art frenzy, producing 16 ladies in landscape pictures (nine Irene's and seven Carole's) and 11 grandchildren pictures, mostly at seaside locations as well as the aircraft pictures.

Part 1
Apprentice

Chapter 1
School

Spring/Summer 1955
Pershore Secondary Modern School

In my last term at school, the activities were centred on careers and finding a job. To this end a career advisor was on hand to guide us. However, I had seen an advert in the Sunday Express for careers as an aircraft artificer in the Fleet Air Arm and sent off for brochures and application forms. I told the career advisor about this during my interview and he decided to leave me to get on with the application.

Later, I was at a desk filling in the application form when my form master, Mr Howell, asked me what I was doing. When I told him, he said, "You don't want to do that, an apprenticeship at RRE Malvern would be a much better bet."

On this advice, I went back to the career advisor to get an application form for apprenticeships at RRE Malvern.

The RRE was the Radar Research Establishment, which had started life at Bawdsey Manor near Felixstowe, under the direction of Sir Robert Watson Watt, the radar pioneer. When the Second World War started the establishment, moved to Dundee, then Worth Maltravers, near Swanage, then Malvern Boys College, before arriving at its final locations at the HMS Duke site in St Andrews Road and the Pale Manor site at Leigh Sinton Road. The establishment gathered a number of names along the way, ending up as the Telecommunications Research Establishment (TRE). Over the years the name changed to Radar Research Establishment, then Royal Radar Establishment after a royal visit in 1956, then Defence Evaluation and Research Agency, (DERA). Today it is part of the technology company QinetiQ.

The recruitment procedure for apprenticeships at RRE included written papers on maths and science and an interview. At the interview, I took along a

Fretwork combination safe money box I had made and the panel seemed quite impressed. Fretwork was a trade name for children's hobby kits which consisted of a set of plans, a sheet of 5 mm plywood and a band saw. I was successful and invited to report to Malvern on 15 September.

My last school photo in the summer of 1955. I am second left in the back row, complete with 'Peaky Blinders' haircut.

Chapter 2
Craft Apprentice

1955–1957
Royal Radar Establishment/
College of Electronics, Malvern

Craft apprenticeship in Instrument Making including;

12 months in the training workshops

6 months with the Pattern Maker

Education: City and Guilds in Machine Shop Engineering

Park View Hostel

On 14 September 1955, I arrived at Park View Hostel, 1, Abbey Road Great Malvern, to take up residence for the next six years. The apprentice's hostel was owned by the Ministry of Supply, the then County Hotel having been commandeered during the war as a staff hostel for the newly relocated Telecommunications Research Establishment. It was however managed by the YMCA.

For the first year, I shared a large room on the first floor with Mike Pearce, who was a fellow student at Pershore Secondary Modern School and Bob Yarnold. The former was called Mac to avoid confusion with me. In fact, there were no less than four Michaels in our intake. I was called Mike and the other two were Mick Clare and 'Rocky' Stone.

Because the hostel was on a steep slope the ground floor at the front was the second floor at the back. As one gained seniority you graduated to first of all a double room and finally a single room on the top floor.

As you passed through the rotary door at the front you were confronted with a hatch on the right housing the receptionist, who apart from the warden, a stocky irascible Welshman, was the only member of staff of British descent. The rest were displaced persons from Eastern Europe with varying commands of English. This is where we booked in and paid our weekly rent.

Next to this was a corridor leading to the ground floor student rooms followed by the entrance to a large lounge furnished with numerous armchairs and round tables. Just inside the door was a large table where sandwiches, tea and coffee were laid out for evening supper. At the end of this long room was a large bay window overlooking a large walled lawn and an alcove containing a record player.

Next came a large notice board and the entrance to a huge spiral staircase which connected to all floors. At the bottom of this stairwell, which was the basement from the front but the ground floor from the back, was a large Coca Cola dispensing machine which delivered the bottle with a loud clatter as it trundled down a chute to the dispensing trough. Some students claimed that they could get high by adding aspirin to the drink, something I never tried myself. In fact, I never saw any drug taking by students apart from alcohol and cigarettes. This was long before the days of pot, ecstasy, coke, acid and speed.

In fact, the college was quite strict in enforcement of expected codes of behaviour. An example of this involved two of the students who got involved in a drunken brawl at a dance at the Winter Gardens and ended up being arrested and spending the night in a police cell. Being reported on in the local paper added to their downfall and this resulted in instant dismissal from the college and their apprenticeship.

Another pair of students, who I seem to remember came from Liverpool, developed an enthusiasm for communism, including a subscription to the newspaper of the British Communist Party, 'The Daily Worker'. When the college management got to hear of this, the pair were summoned and told in no uncertain terms that expressing communist sympathies was incompatible with their signatory to the Official Secrets Act and continuing to do so would result in their dismissal.

In fact, one of them was very good at debating and when during an election campaign a Conservative minister took part in a debate at the Winter Gardens our student 'tore him to shreds' in the debate, much to his embarrassment.

The basement/ground floor also included a ballroom, more of which later, a laundry, a snooker table and a TV room, which was joined by a second one when ITV came along with the new novelty of advertisement breaks every 15 minutes. These were not particularly well attended except when there was a football match on.

There was one broadcast that I never missed. I was one of half a dozen or so students addicted to the Goon Show. Every Monday evening at 8:30 we would gather in the room of a student with a radio, tune into the BBC Home Service and get our weekly fix.

On the left-hand side of the entrance was a large rack of lettered cubby holes for students' mail and the entrance to a large dining room. The spiral staircase was the scene of an annual Christmas ritual where students gathered for a sing song. It started with carol singing from song sheets but eventually degenerated into music hall type songs and bawdy songs. One of these songs I particularly remember was probably a Welsh rugby song. It was sung to the tune of the hymn 'Onward Christian Soldiers' with the words; "Lloyd George knew my father; my father knew Lloyd George" repeated ad-infinitum.

There was even an unofficial college song to the tune of the American folk song 'Red River Valley', the chorus of which ended with the words 'and there you will find an apprentice with his beer cigarettes and a bird', which is the only bit of the song I can remember.

The College of Electronics

The next day we reported to RRE South Site in St Andrews Road, and spent the morning on the initiation procedures including a medical examination, issuing of security passes and signing the Official Secrets Act, before reporting to the College of Electronics. At that time, the south site was dominated by an enormous rotating radar scanner which I seem to remember was called Orange Yeoman.

There were two classes of apprentice in RREs apprenticeship scheme, craft apprentices, including me who were aiming for City and Guilds in Machine Shop Engineering and Electronic Apprentices who were aiming for a Higher National Certificate (HNC) in Electrical Engineering. The former was being trained to fill craftsman posts in one of the establishment workshops and the latter to be employed as a scientific officer in one of the establishment's laboratories.

In 1955, the college was located in a 'H' block in the southeast corner of the establishment. It was a rather dark dingy building. In 1956, the college was moved to another 'H' block on the north site at Leigh Sinton Road. This building had a new two-story office block built on the front making it look more imposing than the old college and the building had a lighter airier feel about it. The front 'office block' consisted of teaching staff offices and classrooms and the 'H' Block consisted of more classrooms, training workshops and laboratories.

Soon after we moved into the new college building there was a royal visit to RRE by the queen and Prince Phillip. The visit included a tour of the college with the royals chatting to us while we pretended to operate machine tools. This was my first of two encounters with members of the royal family. After the visit, the name of the establishment was changed to Royal Radar Establishment, still with the same initials.

The training workshop had a good selection of machine tools with lathes ranging upwards from small Myfords. The milling machines were a more interesting collection including machines commandeered from workshops in Germany and Italy at the end of the war. These included a Maserati and Wanderer milling machines. The former being more famous for their racing and sports cars than machine tools and the latter being part of the Auto Union group whose Ferdinand Porsche designed racing cars, with Mercedes Benz, dominated pre-war motor racing. The rest were mostly Archdale's produced in a factory in Worcester. There was also a small planeing machine and a couple of shapers.

The laboratories also had a useful range of test rigs. The mechanical engineering rigs included tensile test, hardness and modulus of elasticity test rigs. The electrical engineering rigs included a mercury arc rectifier and something called a Wurmshurst machine, which I believe had something to do with static electricity. Both issued 'lightning flashes' when operating which were quite spectacular.

The first-year programme in the training workshop consisted of a set programme of tasks which included making a test piece, a set square, a pair of G clamps and a scribing block. The second-year programme was dedicated to an engineering project; in our case it was to build a large-scale model working steam railway engine.

Pattern Making in North Site Workshops

I was diverted from this for the first six months, because the interview panel had been impressed by my woodwork skills and selected me to do an apprenticeship in pattern making instead of instrument making. I was transferred to the north site workshop under the pattern maker who was to be my apprentice master.

However, in the City and Guilds examinations the following summer I was top of the class and realised that I had a good chance of being 'promoted' to a Student Mechanical Engineering Apprenticeship, leading to a career in Engineering Design and switching from the City and Guilds course to a HNC in Mechanical Engineering so I asked to be reverted to an Instrument making apprenticeship which was reluctantly agreed.

Another reason for my change of heart was that I did not get on with my pattern maker apprentice master Frank, who was an arrogant boastful individual. The woodwork shop had a Wadkin woodwork milling machine which produced enormous quantities of shavings when operating and struck me as rather dangerous, having razor sharp high-speed cutters which were not particularly well guarded. The same applied to the high-speed circular saws.

One activity in the carpenter's shop which did interest me was the manufacture of fibreglass panels. The three carpenters made the moulds used and had been trained in the fibreglass matting and resin layup process. The main item manufactured in fibreglass was radomes for radar scanners.

My main memory of working in the main workshops, then and later, was the BBC radio broadcasts of 'Music While You Work' over the Tannoy System at mid-morning and mid-afternoon. The music was mostly dance bands with the occasional brass band. One I particularly remember was the Big Ben Banjo Band and the sound of massed banjos was distinctly weird. A regular feature of life in the RRE workshops was fire alarm sirens. A number of the employees were volunteer firemen and at the sound of the siren they 'scrambled' into action.

Another thing I remember about the North site main workshop was a first aid man who was very short sighted and wore magnifying spectacles for intricate work and first aid. As a prank, I threaded some fine locking wire through the end of my finger and asked him to remove my 'splinter'. He donned his magnifying spectacles, picked up a pair of tweezers and started to pull the 'splinter' out.

When the 'splinter' had reached a length of several inches, he cottoned on, dropped the tweezers and shouted out, "Go away you idiot."

Another interesting thing about North site was a field to the north of the main workshop which was full of old portable radars which were used in the latter stages of the war after D Day. Some of the radars were captured German units.

College Regalia

What passed for a uniform for the college consisted of a black blazer with the college badge sown on the top pocket and the college scarf and tie. The badge was a circular affair with the Latin inscription 'Caelique Meatus Describent Radio' around the outside. Apart from the last word, I don't think any of us had a clue what this meant or really cared. The rest of the badge consisted of some zig zag lightning strikes and two inner circles with wavy lines. We disrespectfully called it the Mac Fisheries badge, after the chain of fish shops, whose badge it somewhat resembled. All these items could be purchased from the local tailor's shop, either Moss Bros or Burtons, I can't remember which.

Malvern Boys College had a uniform which included a straw boater hat so we decided that we should have a 'tradition' too and came to the conclusion that a bowler hat would do the job nicely. However, the rest of a typical student dress usually consisted of scruffy tight jeans and a large baggy pullover. The college principal was not best pleased.

The cars and motorbikes at Park View Hostel

The cars and motorcycles owned by the students at Park View Hostel were a motley collection. Heading the list was a pair of London taxies, an Austin and a Beardmore, which were 'owned' by syndicates, the memberships of which changed over time. I had a share in one of them for six months in my fourth year, before passing my driving test and purchasing my own first roadworthy car, a 1937 Morris 10.

The BSA 10 hp Saloon

Before that, I had purchased a 1934 BSA 10 hp saloon as a 'restoration project' for the princely sum of £15. It was quite an interesting car which featured a pre-selector gearbox with epicyclical gears. Bill Crump, a friend and colleague had a Riley which also featured a pre-selector gearbox which he did not get on with and decided it was more trouble than it was worth, swapping the Riley for a Ford Anglia/Prefect/Popular of the early 'sit up and beg' variety. After six months of tinkering with the BSA, I decided that I did not have the time or technical knowledge to complete the restoration to roadworthiness and sold it for the price I paid for it.

The London Taxies

One of the taxies featured a single cab at the front for the driver, with an external deck where the front passenger seat would normally be. At one stage, a seat from a spitfire fighter aircraft was fitted to this deck to provide an extra seat, albeit out in the fresh air. On one of our excursions to the Motor Show at Earls Court, we were looking for somewhere to park when we had to brake hard to avoid a car in front. This resulted in the Spitfire seat and its passenger parting company with the decking and ending up on the pavement among a collection of very alarmed pedestrians. The seat and passenger were bundled into the back of the taxi and a hasty getaway was made.

On another occasion, one of the taxies broke down with a dead engine at Cheltenham resulting in a telephone call for help. The other taxi was despatched to the rescue but when it arrived at Cheltenham the brakes had failed. This was overcome by using the near useless handbrake and the brakes of the towed taxi by means of signals from the towing taxi.

Another incident involved an individual called 'Split' Waterman, named after a speedway rider with a spectacular cornering style. He was very short sighted and wore glasses with very thick lenses. With his hair combed straight back, he looked a bit like the jazz pianist Bill Evans. Because of his eyesight and erratic driving style he was not a popular driver with the rest of the syndicate. This view was reinforced when one day he lost control in Malvern Link and crashed into a shop window. To be fair however, the causes of the accident were found to be a front tyre blowout and slippery road due to a recent downpour.

Another interesting car was a BSA 3-wheeler, which unlike the Morgan featured front wheel drive. However, it had the nasty habit of shedding one of its front wheels. This oddball collection was added to by its first brand new car when a new student turned up with a Renault Dauphine purchased by his parents. Needless to say, this did not at first endear him to the rest of us with our 'old bangers'.

The Motorbikes

There was also an interesting collection of motorcycles including a Vincent black 'something or other', and a sporty Velocette, a large and heavy bike, owned by a tall blond student, who looked a bit like a youthful Boris Johnson. He had a tiny girlfriend who was given the nickname 'Mini Mouse' and she learned to ride it and even passed her motorcycle driving test. The sight of her riding with tall boyfriend as pillion was somewhat bizarre.

A student called 'Longstaff' had a powerful Triumph twin which he raced around the Malvern Hills including the tricky Jubilee Drive in his shirtsleeves with no goggles or helmet. He would return to the hostel in a state of high elation with streaming eyes and covered in dead squashed flies. These exotic machines were supplemented by a number of more mundane bikes like BSA Bantam and LE Velocette.

Motorcycle Experiences

My first ride on a motorcycle was on a BSA Gold Star owned by a student from Birmingham. I found it quite exhilarating once I had got the hang of leaning over with the bike instead of against it. He had been taught 'tic tac', the sign language used by racecourse bookmakers to signal 'the odds' to each other, and he taught it to a few of us, enabling us to earn some money at Worcester racecourse.

My second motorcycle experience involved a close friend, Davy Laycock whose Christian name earned him the nickname 'Crocket' as the film and song 'Davy Crocket, king of the wild frontier' were very popular at the time.

He had a small Velocette bike which featured a horizontally opposed twin engine, a bit like a mini-BMW. I was riding pillion with him on this bike across the common between Poolbrook and Guarlford, when negotiating a brow combined with a bend, we found our way blocked by a cow lying in the middle of the road. We could not avoid hitting the cow which bent the front forks of the bike. The cow let out a grunt, got up and wandered off, apparently none the worse for the encounter.

The odd thing is, years after, at a College of Electronics reunion at the Abbey Hotel in the mid-nineties, I mentioned the incident to Davy and he said that he had no memory of it. I could tell from this response that as far as he was concerned it never happened. Did I dream it? It sounds a bit like a dream sequence. Anyway, it is a good little yarn. Davy had a very interesting father who was an organ builder who travelled all over Europe renovating church and cathedral organs.

My third and last motorcycle ride turned out to be something of a nightmare. My parents had a caravan located at Brixham in South Devon and I had planned to join them at Easter to prepare it for the hiring season. A fellow student who had an ancient BSA single cylinder bike offered to take me there as he lived at nearby Modbury and I willingly accepted the offer. The first problem was his spare helmet which was a very loose fit and rattled about on my head causing a severe headache by the time we arrived at Brixham.

En-route, disaster struck in the form of a broken chain at Gloucester, near the Bryant and May's match factory. Luckily there was a motorcycle shop nearby and we purchased a replacement chain. Having fitted the chain, we sat off for Brixham but it then started to rain which continued all the way to Brixham. This experience cured me of motorbikes for the rest of my life.

The Bicycle

My main mode of transport for all but the last nine months of my stay at Park View Hostel was however a bicycle I had purchased with money I had earned by working at Uncle Frank's market garden at Birlingham. My daily rides to work to both north and south sites consisted of a fast freewheeling downhill slalom on the outward journey and a tedious uphill slog on the return journey. I also cycled home to Spetchley at weekends, sometimes via Powick and Worcester and sometimes via Upton on Severn and Pershore.

The Winter Gardens and Cinemas

The back of Park View Hostel overlooked Priory Park which included ornamental gardens and an outdoor swimming pool and just down Grange Road was the Winter Gardens which fronted Priory Park. This complex included a dancehall/concert hall, a café/tea rooms, a terrace overlooking the gardens and at either end of the building two cinemas, one of which doubled as the Festival Theatre.

On Thursdays, which was payday, there was a mass attendance of apprentices, which was a very noisy affair with lots of loud heckling, especially for Hammer horror films. It's a wonder we weren't banned, but I think serious cinema attenders soon learned to avoid Thursdays.

The Pendine Apprentices

In my 1955 craft apprentice intake, there were three from the MoD ranges at Pendine, in South Wales, Brian Hackley, Colin Edmunds and Jackie Eldridge. Because this establishment did not have its own apprentice training facilities this was 'sub contracted' to the College of Electronics and RRE.

There were Pendine apprentices in other years as well and I became on friendly terms with most of them. In fact, the majority of them stayed on in the workshops at RRE Malvern, or in Colin's case RRE Pershore when they completed their apprenticeships, only Brian as far as I am aware returning to Pendine Ranges to work.

One of them had the bright idea of organising an England vs Wales rugby match and because there was not enough real Welshmen to make up the Welsh team, I was invited to play for Wales. I protested that I knew nothing about rugby but was persuaded with the brief 'run forwards and pass the ball back'.

On the appointed day, I managed to avoid contact with the ball for the first twenty minutes or so but eventually someone was stupid enough to pass the ball to me. Before I could leap into action, I was engulfed under a pile of rugby players, one of whom fell on my leg against the knee joint, causing me to be 'invalided' off. This left me limping for a few days.

On another occasion, Brian, Jackie and I went to a pub in Malvern Link for a pint or two. Having had too many pints we were walking back to the hostel across Malvern Link Common with Brian and Jackie singing Welsh rugby songs when suddenly the singing stopped and they disappeared. There followed grunts and cursing from somewhere below where they had fallen into what I later found out was a bomb crater left over from the war. Around fifteen minutes later we were walking along Graham Road when a police car pulled alongside and the policeman wound down the window and asked us what we were doing and where we were going. When we replied that we were returning to Park View Hostel, he responded with 'Oh you lot' and drove off muttering to his colleague.

John Lewis, one of the earlier intake Pendine apprentices, ran a haircutting business from his hostel bedroom in the evenings and weekends and his workbench in the south site main workshops during weekday lunch breaks, charging half the price of the Malvern barber shops.

Also, in the hostel were two Ministry of Works electrical engineering apprentices who were being trained to provide building maintenance support. They both came from Hay on Wye, now famous for its bookshops and annual book festival.

Winter Sports

Another incident I remember occurred during a period of heavy snow. We thought that this was a great opportunity for some winter sports so we decided to go tobogganing on the Malvern Hills. We made a number of sledges from odd bits of timber and set off for the hills.

I shared one of these sledges, a small two-man device, with a fellow apprentice called Colin 'Dixie' Dean. We found a suitable spot on the ridge between the Worcestershire Beacon and North Hill to launch our run. We set off with gathering speed which became somewhat alarming.

Our run was terminated when we crossed a very deep gully preceded by a 'ski jump' embankment, causing Colin, I and the sledge to part company launching into the air. The sledge landed first, then Colin landed across it and I landed on top of Colin rendering him with acute pains in the hip and groin but no broken bones. It was a long painful limp for him, back down the hill to the town and hostel.

Happy Partridge

The apprentices were 'managed' by an individual called Mr Partridge, who was designated The Apprentice Supervisor (today he would no doubt be called a recourse manager). He was a stooped up shrunken individual, a bit like Richard III played by Rowan Atkinson. He had a permanent set scowl on his face which earned him the nickname 'Happy Partridge'. That said, he was very good at his job and helped me a lot with useful advice during my apprenticeship.

In the summer of 1957, having achieved very good results in the City and Guilds examinations, I was selected at an interview to be 'promoted' to a student mechanical engineering apprentice which was geared to eventual employment in a design office.

Park View Hostel as it is today, presumably luxury apartments. My final room was behind one of the windows peeking over the roof.

Chapter 3
Student Apprentice

1957–1961 Royal Radar Establishment/ College of Electronics, Malvern

Student Apprenticeship in Mechanical Engineering

Projects included:

1) Design of printed circuit board artwork from wiring diagrams.

2) Design of a 'swastika' configured chassis for a missile guidance system component to fit in a missile casing and provide maximum access for maintenance and modification.

Education included:

Ordinary and Higher National Certificate in Mechanical Engineering plus endorsements in Principals of Electricity, Maths and Theory of Structures.

I Mech E

During the first year of our new apprenticeship, the five other promoted craft apprentices and I took the final year of the Ordinary National Certificate (ONC) in Mechanical Engineering at the College of Electronics with the Electronic Engineering Apprentices, before transferring to the Worcester Technical College to take a Higher National Certificate (HNC) in Mechanical Engineering.

By this time under the advice, guidance and encouragement of 'Happy Partridge', I had set my sights on obtaining the necessary qualifications to gain membership of the Institution of Mechanical Engineers (I Mech E) and achieve chartered engineer status. One of the endorsements I needed was 'Principles of Electricity' at ONC level. There was a snag in this, I had to gain this qualification before taking my HNC exams and there was insufficient time to take the course.

I got around this with the I Mech E agreement, by reading it up in a book and taking the examination at the I Mech E facility in Birmingham, which I duly passed.

Sloper Jenkins

The HNC course was held in the Victoria Institute, a Victorian building in Worcester city centre, which also housed the county museum. This was part of a scattered complex of buildings of Worcester Technical College, which later moved into a purpose built complex in Deansway.

The lecturer for 'Theory of Machines' was Mr Jenkins, who was quite a character. Because of a serious injury playing rugby he had one shoulder higher than the other, which earned him the nickname 'Sloper Jenkins'. If he thought a student was not paying attention or otherwise annoying him, he would flick a piece of chalk at the offender with the aid of a 12 inch rule. He would usually miss the 'target' by a mile, the chalk bouncing off ceiling, walls or windows but sometimes hitting someone else instead. He spoke to students in a dismissive, disdainful manner, a bit like Basil Fawlty talking to Manuel in Fawlty Towers.

North Site Design Office

The training programme for Design Office Apprentices consisted of alternative six-month spells in a design office and a workshop. My first design office period was in North Site Design Office. At that time, this office was mainly engaged in designing towers for radar units to be located at the recently evacuated Defford Airfield site. These towers somewhat resembled pieces of permanent scaffolding with built in stairways and I was employed on providing detail drawings of brackets and the like.

The leading lite of the office was a larger than life leading draughtsman called Len Smith, who did not take his 'day job' very seriously. His 'main job' as far as he was concerned was being the bandleader and drummer in a local dance band. He had a portable record player next to his drawing board on which he sometimes played his record collection during the lunch break. This collection consisted of music by British dance bands like Ted Heath and Sid Phillips and American swing bands like the Benny Goodman and Tommy Dorsey bands.

When he discovered my enthusiasm for jazz, he invited me to bring some of my records in to play. I chose mainly mainstream jazz by British bands like the Humphrey Lyttleton band and American small groups led by the likes of Buck Clayton, Johnny Hodges and Lester Young as well as the big bands of Duke Ellington and Count Basie. He quite liked the latter, but did not think they were as good as his Benny Goodman and Tommy Dorsey records. When I brought a Charlie Parker record in, he was horrified, declaring that it sounded like a manic turkey gobbling. I wonder what he would have made of Ornette Coleman and Cecil Taylor.

The section leader of the office was a prematurely aged stooping senior draughtsman who walked with a noisy shuffling gait. Len disrespectfully labelled him as HMF or 'his master's feet'. One of the design offices 'customers' was a RAF squadron leader and whenever he made an appearance Len would break out into a tuneless whistling rendition of the march of the royal air force.

Bob, swimming, case hardening, bell ringing and an Austin Seven

Included in the college timetable was a weekly physical exercise session at a gymnasium on North Site, which probably seems anachronistic today and which I did not enjoy very much, not being particularly athletic. This was occasionally replaced during the summer by visits to the swimming pool in the Winter Gardens.

These were usually livened up by Bob Lawrence, a large heavily built individual doing belly flops in an attempt to empty the pool and drown everyone. In his swimming trunks, the skin on the whole of his upper torso was revealed to be severely folded and scarred by what he told us was a boiling water accident when he was a baby.

He was an odd individual who would display sudden bursts of enthusiasm and rush at it like a bull at a gate. He involved me in one of these when we were both in South Site main workshops and getting a bit bored. This was partly because we Design Office apprentices, were not seen by workshops staff as 'proper apprentices' and they were at a loss as to what to do with us. That week we had attended a lecture on heat treatments for steels and Bob decided that we should have a go at case hardening.

To this end we went into a small alcove forge area to the side of the machine shop with some bits of scrap steel and carbon powder. We then set to with a blow torch trying to emulate what we remembered of the process. What we hadn't noticed was that we were creating huge clouds of thick black smoke which were wafting up to the ceiling of the alcove and out into the workshop, causing alarm and consternation, followed by extreme annoyance by the supervisors. We were being angrily harangued by one of the latter when Bob came over all hurt innocence and calmly explained that we were carrying out experiments in case hardening. I sometimes think that Bob was motivated to causing as much mayhem as possible in these escapades.

He owned a very early, somewhat decrepit Austin Seven and on one occasion asked me to help him find and cure a persistent misfire and lack of power. In pursuit of this, we poked and prodded about with neither of us having a clue what we were doing. Bob's dad soon lost patience with this and hired a mechanic to fix the problem. Even with my limited knowledge then of automobile engineering, I was amazed at how simple and basic, crude even, that the Austin Sevens mechanicals were. For instance, the fuel system consisted of just a small fuel tank located at the scuttle which gravity fed the updraft carburettor.

Bob's father was a member of the clergy at Malvern Abbey and Bob was an occasional member of the bell ringing team. One day he invited a few of us to a bell ringing session. I soon found that bell ringing was more complicated than I thought and found it a bit alarming with visions of me hanging on to a bell rope fast disappearing into a hole in the very high ceiling.

Visits

We made quite a lot of 'educational' visits by coach to engineering organisations in the Midlands and beyond. These included, the Bristol Aircraft Company at Filton near Bristol, then producing the Britannia turboprop airliner, Hawker Aircraft at Bitteswell near Coventry, then producing Sea Hawk naval fighters, British Rail workshops at Swindon and the Triumph Car Factory at Coventry, then producing Triumph Heralds.

Rag Weeks

The highlight of the College of Electronics year for the students was the 'Rag Week' during the late summer where we spent the week doing silly things for local charities. The week culminated on the Saturday with a carnival procession followed in the evening by the Carnival Ball at the Winter Gardens featuring the popular bands of the time such as Chris Barber, Acker Bilk, Humphrey Lyttleton and Johnny Dankworth.

Fire Eaters

For my first carnival procession, I and a few others dressed up as 'Indian Fakirs' with baggy bright coloured trousers and our face and upper torso 'blacked up' with brown boot polish. The 'fire eating' consisted of taking a mouthful of paraffin and squirting it out as a fine mist while igniting it with a flaming torch. The resulting column of flame was very spectacular, but at the cost of singed eyebrows and lips and ulcers in the mouth.

The Shabby Hotel and Stonehenge

The following year, I took part on a float which was made up as the interior of a ramshackle hotel with students and girlfriends dressed up as chefs, French maids, pompous managers and dishevelled customers dressed in pyjamas and nighties. There was a large sign over the float announcing it as the 'Shabby Hotel', an obvious reference to the Abbey Hotel, Malvern's poshest hotel. The hotel did not see the joke and was, in fact, outraged by this and threatened to sue the college. In the end, we got away with an apology in the Malvern Gazette.

Another float I got involved in was a representation of Stonehenge accompanied by druids and cavemen doing a war dance. English Heritage would be appalled at our idea of what Stonehenge was all about. The henge monument was constructed from spare wardrobes stored in the hostel.

The Snowman

Another float, the theme of which I do not recall, decided to feature snowball fights using flour bombs instead. Unfortunately, when the snowball fighters spotted a policeman on traffic duty at a road junction, they decided to turn their attention on him, resulting in him looking like a snowman. I did not witness this incident but my parents, who were watching the carnival procession from nearby did and couldn't believe their eyes to see a policeman being pelted with flour bombs. The carnival organising committee must have been annoyed at this outbreak of hooliganism and duly compensated the constable for damage inflicted.

The Nuclear Beetle

For my final participation in the Rag Week Carnival Procession in 1959, I teamed up with three fellow engineering apprentices and mates, Dennis Underhill, Bill Crump and Pete Bellamy to produce a 'Nuclear Beetle'. We constructed this on a pallet mounted on four industrial castors with a bench seat fitted. This was covered in a shell of wire netting covered in a black drape. The result looked more like a cat's toy mouse than a beetle.

During the building phase one of the others, I think it was Pete, turned up with a damaged wooden barrel of cider which he claimed really did fall off the back of a lorry. We decided that we needed to drink the cider before it all leaked away. To this end we obtained some paper cups and drank the lot. Needless to say, no useful work was done on the beetle that evening. Manoeuvring the barrel to achieve a controlled steady flow of cider from the bung hole proved to be something of a challenge and our first attempts were decidedly messy with cider ejaculating all over the place.

During the carnival procession each of us took turns to sit inside the Beetle lighting and dropping 'jumping jack' fireworks while the other three dressed in black cloaks, hoods and masks towed the beetle. When it came to Denis' turn inside the Beetle, we hooked it up to one of the float lorries.

What we did not notice was the exhaust of the lorry was directed at the Beetle. What we did notice after a time was that the fireworks emerging from the Beetle ceased.

We shouted, "Are you alright?"

Which received no response. We then unhooked the beetle and tipped over the shell to reveal Dennis in a state of collapse and a pale grey pallor due to the exhaust fumes. Fortunately, a St Johns Ambulance crew was on hand to administer first aid and Dennis soon recovered.

During the carnival dance we received a prize of a crate of beer for the Beetle as the best walking entry in the carnival procession. However, the Beetle was later in the evening 'stolen' by a group of 'squaddies' from the nearby Blackmoor Army Camp which is now the site of the Three Counties Show ground. They then promptly issued a 'ransom' demand. This was paid by the carnival committee, but to the local charities not the squaddies, and the Beetle was returned to be dismantled.

The Rag Magazine

A single edition was produced each Rag Week with a mixture of silly stories, poems, cartoons and articles. One I remember was a pseudo advertisement. At the time, there was a ladies' hair treatment called Toni which carried an advert in magazines and newspapers featuring a picture of a pair of twins with the headline: 'Which twin has the Toni set'. At this time, padded bras were available for girls less well-endowed bosom-wise and were known as 'falsies'. Our advert featured a pair of 'top heavy' twins with the headline: "Which twin has the Foni pair?"

The Band

Park View Hostel had a large ballroom on the ground floor at the back, a legacy from the pre-war days when it was the county hotel. During the summer we organised monthly dances inviting local girls and students from local girls' teacher training colleges at Hereford and Shenstone, near Kidderminster.

A group of 'wannabe' musicians including me decided that it would be a great idea to form a band rather than rely on records, so a band was formed with me on banjo, Pete Todd on trumpet, Dave Goody on trombone, 'Split' Waterman on drums, Steve Barratt on piano and later, a clarinet player whose name I can't remember.

We agreed that we would play New Orleans/Dixieland Jazz which was very popular at the time with bands like Humphry Lyttleton, Chris Barber and Acker

Bilk. The band was dominated by Pete Todd who favoured the music of Kid Ory and Teddy Buckner. Unfortunately, we did not get beyond the rehearsal stage before we disbanded, not through animosity or disagreements but disheartened at our lack of competence on our chosen instruments.

A friend of mine, John Watkins, fared much better as a musician. He formed a skiffle group with a bunch of friends and evolved into a fairly competent guitar and banjo player. The group were quite popular at local dances. John later joined a Worcester jazz band who called themselves the Severn City Jazz Band.

The Sick Bay, the Faith Healer and Pentecostal Tabernacle

One of the establishment facilities I became very grateful for was the Sick Bay. It was located in Pickersleigh Road, quite close to the Morgan Motor Works and was like a mini hospital with full-time nurses and a small ward with half a dozen beds. On a Maundy Thursday, I was working at a drawing board in the main design office when I suddenly threw up over it and turned a deathly shade of pale. This was spotted by a colleague; an ambulance was called and I ended up in the sick bay. A doctor turned up later that morning and diagnosed my condition as gastro-enteritis and 'sentenced' me to three or four days in the sick bay. I consequently spent the whole of the Easter Holiday there, very well looked after by the nurses.

The apprentices who were resident at Park View Hostel were allocated to a surgery just down the road from the hostel. The doctor who treated sick apprentices was a rather archaic individual who had an almost life size picture of Jesus holding a lantern in his consulting room. His consultations with wayward apprentices often included a lecture on leading a more upright/moral life and this resulted in us designating him as the faith healer.

On the subject of religion in those days, application forms and the like usually asked one to state your religion. Dennis decided that we should all be 'Pentecostal Tabernacle' and after that we entered this as our religion on such forms.

Lunchtimes

Because we always had a hearty breakfast and dinner at Park View Hostel, I was never particularly hungry at lunchtimes, so unless it was raining or very cold my usual lunchtime activity consisted of purchasing a wagon wheel biscuit at the canteen and going for a long walk over the Malvern Hills or nearby commons.

Cheltenham

Another close friend, Cliff Canning, had a Lambretta scooter and lived at Prestbury, near Cheltenham with a large family of younger siblings. We were both motor racing enthusiasts and we argued constantly over the merits of Mike Hawthorn and Stirling Moss. I favoured the former and Cliff the latter. Cheltenham's most celebrated residents are both musicians, the composer Gustav Holst and the founder member of the Rolling Stones, Brian Jones.

Cliff invited me to spend a weekend at Cheltenham and sample the night life. This consisted of touring around the centre of Cheltenham on Cliffs Lambretta scooter and visiting coffee bars. One of these had replica coffins as seats and tables and because I could not be accommodated in Cliff's house, we arranged for me to sleep in this coffee bar on the Saturday night. When I told Mum that I planned to sleep in a coffin, she became alarmed and arranged instead for me to sleep at the flat of Dad's uncle Horace and Aunty Maggie which was also at Prestbury.

The Von Tripps Committee

At the hostel, there was a committee which organised excursions to places of entertainment. It was called the Trips Committee and I went on a couple of their excursions. The first was Louis Armstrong and his All Stars at Birmingham Town Hall and the second Jazz at the Philharmonic at the De Montfort Hall, in Leicester.

At one stage, Cliff Mac and I were invited to run the committee. Because of our strong emphasis on excursions to motor racing events, we called it the Von Tripps Committee, after Wolfgang Von Tripps, a popular German racing driver of the time, who should have and probably would have been world champion in 1961 but for his untimely death in a crash at Monza.

London

Trips to London were quite popular, usually either to the Motor Show at Earls Court or to jazz clubs. We usually used the London taxies for this purpose but sometimes went by train or hired minibus. The most popular venue in central London was the Cy Laurie Jazz Club in Great Windmill Street near to the Windmill Theatre, famous for its naked lady tableau shows. Cy Laurie was a clarinet player who was a devotee of the 1920s New Orleans clarinettist Johnny Dodds and his band played jazz in the style of bands like the Louis Armstrong Hot Fives and Sevens, Jelly Roll Moreton's Red-Hot Peppers and the various bands led by Johnny Dodds.

We also went on a couple of occasions to Eel Pie Island, on the Thames near Twickenham. The Eel Pie Island Hotel had a ballroom which was the venue for a jazz club and we went there to see Ken Colyer's Jazzmen on the first occasion and the Sandy Brown Band on the second. The air of rundown squalor, the bouncing ballroom floor and entwined couples all over the grounds were all present and correct.

Ken Colyer's Jazzmen made a record of one of their concerts at Eel Pie Island and I much later purchased a reissue of it on CD but there was no sign of recording activity at the one I attended.

The Sandy Brown concert was a chaotic affair and after the band had played for about an hour Sandy announced that they were going to take a ten-minute break. Nearer to an hour had elapsed when the band returned with Sandy pretty tipsy by then and by way of an introduction to their first number, he launched into a long rambling monologue on an obscure topic of no interest to anyone except himself. Some members of the audience soon lost patience with this with shouts of 'Shut up and play some music'.

I got the impression that Sandy's reception was somewhat lukewarm compared with what I remembered for Ken Colyer on my previous visit. He had a very quirky and original phrasing in his playing and he played a lot of his own compositions rather than relying on classic New Orleans repertoire. In fact, I think he was the best and most original jazz clarinet player in the country then and has not been surpassed since.

Sandy had a day job as a renowned acoustic architect and he designed a number of concert halls and recording studios.

Brussels

In 1958, the World Fair was held at Brussels and the YMCA organised holidays at their hostel in the centre of Brussels. I and a group of half a dozen other apprentices decided to take up this offer as the price was very competitive.

On the due date, we set off by train to London and onward to Dover to catch the ferry to Ostend. In Belgium, we caught the train to Brussels via Bruges and Ghent. On arrival at the Gare du Midi, we walked to the hostel which was not that far away. The next morning, we were dismayed to find that breakfast was a continental one of croissants, butter and jam. However, we soon discovered that the nearby Woolworths store had a cafeteria at the back which served huge cream cakes and this became our breakfast supplement for the rest of the week. The hostel did provide a reasonable evening meal and packed lunches if required. However, our daily lunch soon became 'frites' (French for chips) which could be purchased in kiosks served up in paper cones.

There are two races in Belgium; the Flemish, who speak Dutch and the Walloons who speak French. Consequently, all signs were duplicated in the two languages, a bit like being in Wales.

The first couple of days were spent exploring the centre of Brussels including the Manneken-pis, which was a statue of a little boy weeing, and the Cathedral des Sts-Michel-et-Gudule which had beautiful stained-glass windows in the domed choir. Here one of our party whose real name I can't remember but who was known to one and all as 'birdbrain', not because of any lack of mental capacity but because his head in profile resembled a budgie, lay on his back on the floor under the dome taking photos.

Another of our party, Ken Nosworthy, who came from Devon and spoke in a soft west country accent, a bit like the celebrity chef Rick Stein, was the only one of us with any competence in French. He became our spokesman and when we needed to make transactions in shops or cafés he would speak in badly mispronounced French in a very loud North Country accent, a bit like a Monty Python sketch. This would leave the recipient baffled and the response was often 'Are you English', followed by the rest of the conversation being carried out in English.

The Undone Jeans

Towards the end of the holiday we visited a large clothes shop and I purchased a pair of powder blue jeans with leather beading down the outside leg seams. I thought they were pretty cool and wore them continuously. They were fine until about three or four weeks later and back in the UK, the seams started to come undone due to rotten cotton. Mum had several goes at repairing them, but in the end, we gave up and they ended up in the rag bag.

The World Fair and a Belgian Family

We spent a day at the World Fair and the adjacent Bruparck giant funfair. The main attraction for me was the Atomium giant metal sculpture which had an incredibly fast lift to the top sphere which had spectacular views of Brussels and district. The USA enclosure was also an attraction for us with a lot of new technology on display. One exhibit I particularly remember was a stereo HiFi system featuring a pair of large speakers in spherical enclosures. Stereo was so new then that I don't think it was available to the public in the UK.

We then decided to try the rides in the funfair and met a group of girls some of whom spoke good English. They spent the rest of the afternoon with us: I think they wanted to practice their English. The next morning two of them, sisters, turned up at the hostel with an invitation to Sunday lunch with their family. On the Sunday, the father picked us up from the hostel in his car. He spoke some English but was not as fluent as his daughters.

The conversations over lunch centred on him asking questions about us including where we lived. When we answered Malvern, he shook his head enquiringly, we then said Worcester which also elicited a puzzled response. When we responded 'near Birmingham', his face lit up.

"During the relief of Brussels in the war I befriended a soldier called Joe Brown who came from Birmingham, do you know him?" he responded.

He obviously had no idea of the size of Birmingham. He also told us that during the occupation Brussels was notorious for collaboration, something he was rather ashamed of. I wonder if this was anything to do with animosity between Walloons and Flemish.

Excursions

We then decided to look beyond Brussels and in our first attempt three of us decided to hitch hike to Antwerp. We took a tram to the outskirts and set out walking along the main road to Antwerp. We had been walking for about an hour with no success when a large black Mercedes Benz stopped and picked us up. The driver turned out to be a Brazilian diplomat who seemed quite intrigued by three engineering students from England. For the return journey, we decided we had had enough of hitch hiking and took the train. The second-class carriage featured seats with wooden slats like a park bench which were very uncomfortable.

The next two days we took coach excursions to Luxembourg and the Battlefield of Waterloo. The latter was a flat featureless landscape, except for a large artificial mound with a statue of a lion on top. Near Hugemont Farm where the fiercest fighting took place was a building with a large panoramic painting of the battle in progress.

For our night-time entertainment, we found a nearby coffee bar which featured a long narrow area with seats down one side and a walkway down the other. At the back was a jukebox which played loud American rock and roll music and a small area for dancing. This featured a floor of glass blocks with lights underneath. This enhanced the spectacle of the twirling skirts of the jiving girls. Returning from evening entertainments we would often march along whistling 'Colonel Bogey', a song popularised by the film 'Bridge over the River Kwai', just to let people know we were Brits.

The week came to a close and with it the return to Malvern. The ferry crossing was spent on the rear deck, in glorious sunshine in the company of a group of returning girls.

The Foundry

The establishment did have a small foundry in the South site main workshops but it was very infrequently used and most casting tasks were contracted out. It was decided therefore that we needed to gain experience of foundry techniques and the six of us were despatched to a foundry at Willenhall, between Wolverhampton and Walsall for eight weeks in the summer of 1959. The firm was called C&L Hill and they were part of the Rubery Owen Group more famous for BRM racing cars.

They specialised in aluminium castings and these included sand castings, die castings and an intermediate technique of multiple small castings from 'biscuit' moulds of 'fired' sand. The firm's buildings consisted of a three-storey front office block which was like a film set because behind, by stark contrast, were a series of large dilapidated corrugated iron sheds with dirt floors.

For the duration of our posting, my colleagues stayed at a B&B at the south end of Wolverhampton, but I stayed with Aunty Molly and Uncle Jack who lived in a large house which included a tobacconist and sweet shop at the front, situated on Summer Hill, at the crossroads of Kingswinford, south of Wolverhampton.

The staff of the foundry had a high percentage of coloured workers with Asians in the skilled jobs and West Indians in the labouring jobs. The foreman was a fairly tough unapproachable individual who wore a flat cap. There were examples of graffiti on the walls accusing him of being a 'nigger lover'. One individual was a pattern maker who made metal patterns for die castings. He smoked continuously and when the cigarette got so short that it burned his lips, he took another out, lit it from the stub and replaced it. The cigarette never left the corner of his mouth and he had a brown nicotine stain starting at the corner of his mouth, extending upwards past his nose and up to his hairline and the peak of his flat cap.

Football, Boating and Roller Skating

Our evening leisure activities included a couple of visits to Molineux, the Wolverhampton Wanderers football ground. In the fifties, the 'Wolves' were one of the top teams in the country and used a tactic of long passes which avoided too much running about and early fatigue. The captain was Billy Wright, probably the first celebrity footballer who was married to one of the Beverly sisters, a very popular singing group of those days.

We also visited West Park a few times hiring boats on the boating lake. On one occasion, we started an oar splashing tournament among our three boats which everyone else on the lake soon joined in.

On another occasion, we visited a roller-skating ring and I was trundling slowly and gingerly around the outside when a group of girls asked me to join them. I ended up on the outside of the chain of girls travelling at nerve shattering speed around the outside of the ring.

Aunty Molly and Uncle Jack

My colleagues used to go home for the weekends but I stayed with Aunty Molly and Uncle Jack, where I was spoiled rotten. I joined in some of their social activities including a couple of visits to a local pub, which had suffered from mining subsidence. The brewery who owned the pub instead of having it demolished had it shored up and added to the now uneven floors and walls by refitting windows, doors and furniture at odd angles. In the centre of the bar was a large table, which if you placed a bottle on its side would appear to roll uphill.

Uncle Jack was a glazier by trade and he had acquired the skill of leaded lights during his apprenticeship. This was an art deco version of the medieval picture church windows which was very popular on new suburban houses in the thirties and forties. During the Christmas/new year holiday just after the second world war the managing director of Boulton Paul Aircraft Ltd, which was based at Pendeford Airfield, north west of Wolverhampton, suffered a broken leaded light panel on the front door of his house. He was having great difficulty in finding a glazier who was skilled in leaded lights and willing to work over the holiday when someone told him about Uncle Jack, who was duly contacted and did the repair to the delight of the customer, who became on good terms with him during the repair.

So pleased, in fact, that he set Uncle Jack up in his own glazing business by giving him a van, finding him a business premises and providing him with work on the factory site. Uncle Jack was a very astute businessman and his business was thriving by the time I got to know him. He was always on the lookout for new business opportunities and the only time this did not work out was the purchase of a cherry orchard. I am not sure of exactly what went wrong, successive years of bad harvests or difficulty in recruiting pickers perhaps.

There was a British Legion Club across the road from the shop and Uncle Jack was a member. We used to go there for a pint and a game of snooker. We were both fairly useless at the game but were usually lucky when playing games with other members, which was met with a mixture of amusement and annoyance. One of the members was the Worcestershire cricket player Jack Flavell.

43

At a charity harvest festival function, Uncle Jack contributed a lot of Chrystal glassware for raffle prizes. He was on friendly terms with some of the managers at the nearby Stuart Crystal factory and used to buy 'seconds' in bulk at much reduced prices to sell in the shop. He bought a very large number of raffle tickets and was continually winning the prizes he had donated much to everyone's amusement.

At this time, Uncle Jack had an interesting car. To all appearances it was a Jaguar Mk II saloon like the one driven by Inspector Morse in the TV series, even the same colour. Except that it was not a Jaguar but a Daimler. Instead of the Jaguar light alloy double ohc straight six engine, it had a 2.5 litre Daimler V8 engine from the Daimler SP250 sports car. Externally the only difference was a fluted top to the radiator grill.

My cousins, Ann and Jean, still lived at home then but I did not see much of them because they were both courting. That is except for Ann who would come into my bedroom at around ten o' clock after a date and regale me with the problems she was having with her boyfriend when all I wanted to do was go to sleep. She later broke off with Ray, a trainee accountant and took up with David, a loud and boisterous but very likable character who she eventually married.

Aunty Molly had a small black poodle and on the occasional evening we would take it for a walk on Kinver Edge, an Iron Age hill fort with splendid views over the surrounding countryside. Nearby were some cave dwellings carved out of the local Devonian sandstone which were occupied until the thirties.

There was a large pub to the south of Kingswindford called the Waggon and Horses. It had a large bar which was converted into a ballroom by the simple expedient of moving the tables and chairs to the outside for dances on Thursday evenings. Towards the end of our stay Aunty Molly decided we should go there with my colleagues. It turned out to be a lively and enjoyable evening and I had a couple of dances with a girl with a striking yellow dress after some encouragement from Aunty Molly.

Welding and Modern Art

On our return from the foundry, we were assigned to a brief training session on welding at the college workshops, which we were sent on, two at a time. Pete Bellamy and I were the first to do this and were given a good briefing on the use and safety of their newly acquired arc welding equipment which took up the first morning. We were then left to our own devices to practice arc welding over the next day and a half.

At first, we were at a loss as to what to do with the arc welding kit and a pile of scrap steel sheet and plate put at our disposal, but decided we would produce a modern art sculpture. To this end we started out with a fairly large piece of steel plate as a base and started to weld oddly shaped bits of steel sheet to it and each other in a generally upward direction. After a day and a half of this, we ended up with a large grotesque monstrosity reminiscent of a Vorticist painting by Wyndham Lewis or a cubist painting by Picasso which left the instructor duly unimpressed. It probably ended up in the scrap bin rather than the Tate Gallery.

Carole Mary Everley

In the late autumn of 1959, I met Carole at a Friday free session at the Victor Sylvester Dance Studio above the Gaumont Cinema in Worcester's Foregate Street to start four years of courtship. This took care of most of my leisure time over this period and our activities ranged over the Worcester, Malvern and Spetchley areas.

Our entertainments included pubs, coffee bars, which were all the rage then, the new phenomenon of Chinese and Indian retardants, a jazz club, dances at the Worcester Corn Market and Malvern Winter Gardens and concerts at the Gaumont Cinema. Apart from our first date we did not go to the cinema, unlike most courting couples at that time.

After I had passed my driving test and purchased a 1937 Morris 10, the longer excursions, we could now indulge in included a couple of visits to Wales and a few visits to the Wiltshire/Somerset border where Carole's grandparents lived. We also visited Shelsey Walsh, Prescott, Silverstone and Goodwood for motorsport events.

After about 18 months, I became dissatisfied with the slow, heavy, cumbersome Morris 10 and started looking for a more agile, modern (post war) car and came across a 1953 Renault 750 which I purchased from a policeman. This car was a UK version of the Renault 4CV which was assembled at the Renault factory in Acton, West London.

Coach Trips

Our leisure activities also included the occasional coach trip. These were very popular at the time, especially trips to the seaside. The only snag was that the nearest resort, Weston Super Mare, was 100 miles from Worcester. It was at Weston Super Mare where I had my only equestrian experience riding a donkey along the sands at about five years old.

Two other popular resorts were Barry Island and Porthcawl in South Wales. These were resorts developed to cater for the needs of the workers of the coalfields of the Rhonda Valley and the steelworks of Port Talbot and then consisted mainly of funfairs, amusement arcades, pubs and fish and chip shops so did not particularly appeal to us.

Another popular trip was called a mystery tour, which was in reality a pub crawl. These were popular with the men but less so with wives and girlfriends. We went on at least one trip to Weston Super Mare but our favourite was a trip to Barmouth, a resort in Cardigan Bay in mid-Wales. The journey included spectacular mountain scenery between Machynlleth and Dolgellau, and Barmouth itself was a very nice resort.

The day started out as sunny but by the time we arrived at Barmouth had deteriorated to overcast and windy. In spite of this, Carole was not going to spend a day at the seaside without a paddle in the sea. I still have some photographs of a windswept Carole paddling in the sea looking pensive. For these, I asked her to wander about while I did my David Bailey impression.

Semiconductors and PCBs

I spent some time in the Electrical Design Office which I found interesting and stimulating. The section leader took great pains to give me interesting tasks and the main one was designing printed circuit layouts from wiring diagrams for the tracers to produce ink artwork for the printed circuit boards (PCBs). These and the use of semiconductors in place of thermionic valves was new technology then. These PCBs were to be used in the establishments' main computer which filled a huge room and probably had less capacity than a modern smartphone.

The Swastika Boxes

One of the more interesting tasks I had in the main mechanical design office was a chassis to house the components of a missile guidance system. These were to fit inside the missile casing which was a 20 cm aluminium alloy tube. I came up with a swastika configuration in which the 'arms' were able to fold out to gain access to the components for testing. These guidance systems and their missiles were test fired at the MoD ranges at Larkhill on Salisbury Plain.

National Service

During the period of my apprenticeship National Service was in place. This involved males of 18 years old being conscripted into the armed services for three years. Those servicing apprenticeships had this postponed until completion of the apprenticeship.

As I approached the end of my apprenticeship, I was looking forward to this with a certain amount of trepidation and was relieved when it was abandoned by the government six months or so before the end of my apprenticeship.

I later learned that I may have missed out on a useful experience as several of my older colleagues had a rewarding and interesting service, including postings to interesting parts of the world because of the qualifications and experience they gained during their apprenticeship.

The band. From left to right, Dave Goody, 'Split' Waterman, Pete Todd, me and Steve Barratt.

The four Michaels plus one at the Winter Gardens, Malvern in August 1958. From left to right, 'Rocky' Stone, me, Mick Clare, 'Mac' Pearce with giant cigar, purchased in Brussels and Charley Tarling.

Barmouth
1960

The Carole Barmouth 'David Bailey' photos

Part 2
Design Draughtsman

Chapter 4
Test Chambers

March – October 1961
Royal Radar Establishment, Malvern

Design Draughtsman

Design studies for:

1) A dust chamber.

2) A corrosive atmosphere chamber.

Design Studies

On completion of my apprenticeship on 24 March 1961, I was taken on as a design draughtsman in E Block Design Office, on RRE South Site. This was followed by the successful completion of a Higher National Certificate in Mechanical Engineering the following June. The following September I enrolled for the Theory of Structures course at Worcester Technical College to complete the Institution of Mechanical Engineers part two academic requirements,

I was allocated to a leading draughtsman and we were assigned a project to develop two environment test facilities, a dust chamber and an industrial atmosphere chamber. Having established with the customer the size of the equipment to be tested and therefore the size of chambers required we embarked on design studies.

We decided that the dust chamber was likely to be the easiest to accomplish so we started that one first. The main feature was a method of distributing the dust and we looked at blowers and fans before deciding on a system of a pair of paddles at the base of the chamber. To see if this method gave a reasonable distribution of dust, we constructed a scale model from balsa wood with the paddles driven by small electric motors.

We did not spend much time or effort on the industrial atmosphere chamber. Our first problem was to get the customer to define what kinds of atmosphere he wanted to simulate. He could only come up with a salt water spray of a seaborne environment and oil/petrol fumes of a generator. The other thing we considered was the materials to use in the construction of the chamber and pipework, including stainless steel and glass.

Indoor Model Aircraft

The leading draughtsman who was my mentor had an interesting hobby. He made and flew indoor model aeroplanes. The location where he flew his models was the Gaumont cinema in Worcester. The construction of these models was totally aimed at minimum weight and consisted of a minimal balsa wood space frame covered with a skin made by pouring the ingredients onto a bath of still water. Propulsion was provided by a large in proportion propeller driven by a wound-up elastic band. One needed great care in winding up the band because if it was over wound, it would cause the frail frame to collapse. As well as fixed wing models, he also made and flew helicopters and ornithopters with flapping wings like a bird.

When it came to hobbies, there was a very useful facility at RRE Malvern. It was the Component Recovery Stores, where no longer required components were stored and available to employees free of charge. Having chosen what you wanted to withdraw, you were issued with a permit listing the items acquired to show the MoD Police gate guard when you left the establishment with them. I never made use of this facility myself but some of my electronic apprentice friends did, for making radio receivers, audio amplifiers and the like.

The Move to Pershore

One of my fellow apprentices, Dennis had been allocated to the RRE airfield at Pershore and he enthused about the interesting projects he was working on and encouraged me to join him there. I therefore requested a transfer which was agreed and I started work there in October 1961.

Chapter 5
Aircraft Installations

1961–1965 Royal Radar Establishment – Pershore Airfield

Design Draughtsman
Projects included:

1. The design of the mountings for a prototype sideways looking radar (SLR) in the bomb bay of a Canberra aircraft.
2. The design of the installation of a prototype optical linescan mapping system in a Canberra aircraft.

First project

On arrival at the Design Office at RRE Pershore, I was placed on the team of a Senior Draughtsman called Harry Cameron, who originated from Ulster. Judging by his name, he was from the protestant/unionist community but we never discussed his Ulster background or his political views.

The first job he gave me was the installation of a small cruciform aerial in the upper equipment bay hatch, behind the cockpit on a Canberra aircraft. I achieved this by designing a shallow pyramid structure in 18swg L72 aluminium alloy. Most of the structures we designed at Pershore used either L72 or T45 alloy steel tubing.

The Aircraft Fleet and The Airfield

At the time I was there, the aircraft fleet consisted of half a dozen Canberra's, a pair of Varsities, a Viscount, a Hermes, a Hastings and Whirlwind and Wessex helicopters. A Viscount V bomber was later added to the fleet to have a number of prototype new radar and mapping systems installed including a rear warning radar. Unfortunately, on its maiden flight after the installations it crash landed on take-off. This resulted in a replacement Viscount being obtained and the installations being repeated with a considerable delay in the programme.

Between 1940 and 1942 when TRE was at Worth Maltravers the supporting air fleet was located at Christchurch Airfield. With the move of TRE to Malvern, the air fleet was relocated to Defford, an airfield four miles east of Malvern, However, by the mid-fifties the limited runways at Defford were deemed inadequate for the new generation of jet military aircraft like the Canberra and V bombers so was relocated to Tilesford Airfield, two miles northwest of Pershore, which was renamed RRE Pershore.

At the time of the move, the main runway was extended and the church spire of St Egburta's Church at the nearby hamlet of Abberton was dismantled and removed for flight safety reasons. The masonry was numbered and stored at RRE Pershore with the intention of restoring it if/when the airfield was closed. However, when the airfield was closed in 1977 it was deemed impossible to restore it in a structurally safe manner, costs and utilisation probably came into the equation too.

Sideways Looking Radar

On completion of the cruciform aerial installation, I then joined the rest of Harry's team on the design of the installation of a prototype sideways looking radar (SLR) in the bomb bay of a Canberra. This was a terrain mapping system being developed for use in the multi-role TSR2 aircraft. I believe the overall concept of the installation was Harry's and it was a very good one.

For this major installation, the bomb doors were partially cut away and a representation of the under fuselage of the TSR2 was inserted. The mounting structure for the sideways looking radar consisted of an 'I' beam constructed from 18swg L72 aluminium alloy which ran the length of the bomb bay. This was attached to the bomb beam mounting points in the roof of the bomb bay by a space frame structure which was my main contribution to the project.

I also designed another space frame structure to mount a very large cylindrical transponder unit in a Varsity aircraft and was also involved in the installation of the optical line scan, another mapping device being developed for TSR2. With the cancellation of TSR2, both SLR and line scan were further developed and ended up in underwing pods fitted to Phantom and Tornado aircraft.

Colin Chapman and the Lotus Seven

My designs at Pershore were influenced by what I knew of the work of Colin Chapman on his Lotus racing cars. His earlier racing cars chassis were designed around purist space frame principles, but in 1962 he came up with the revolutionary Lotus 25 which employed a unit construction body/cassis/fuel tank assembly in aluminium alloy sheet. This car dominated Formula 1 for the next 4 years.

At this time, there was a competition in Autosport magazine with the prize of a Lotus Seven. To win this prize, one had to estimate the maximum speed and acceleration curve of the car. I decided to have a go at this and to this end looked up the technical specification of the car from sales brochures and motoring magazine articles. There was no information on drag coefficient or frontal area so I made an estimate of these and stated the reasoning behind these estimates.

I was working on this at my desk when I was spotted by the installation manager, Tom David. Instead of reprimanding me he surprisingly took a keen interest in what I was doing. Incidentally, I did not win the Lotus Seven.

The Stress Office

I always carried out stress calculations for my designs using the knowledge gained on the theory of structures module of the HNC course, I was the only designer in the office to do this and it was soon spotted by the chief engineer, resulting in me being invited to transfer to the Stress Office. There were two regimes used in carrying out stress calculations. The first was the loads imposed by the flight envelope of the type of aircraft concerned plus a safety factor. However, if the installation shared a space with aircrew or fight test engineers the 'crash case' was imposed, which used a deceleration of 15 g as the design criteria.

The Stress Office post did not work out well for me however as I found that analysing the efforts of colleagues less satisfying than the creative activity of design. Also, I did not get on well with the senior Strassman who was a somewhat strange individual of East European origin who came to Pershore from the Gloster Aircraft Company. I therefore asked to be transferred back to the Design Office which was agreed with some disappointment by the Chief Engineer.

David Henderson

Our Chief Engineer, David Henderson, was a very dynamic hyper active individual who had previously been Chief Designer and therefore had an extensive knowledge of aircraft structures and modifications. He was well liked and respected by all his staff in spite of his somewhat erasable manner and was the only boss in my career who I actually admired. However, towards the end of my period at Pershore he started to suffer from ill health and became virtually blind.

The Errant Leading Designer

One of the leading designers, Norman by name, was notorious for never producing any drawings. If he could not get an assistant to do this for him, he would discuss with the workshop foreman and craftsmen what he wanted and cover the documentation with a scantily written Design Office Instruction (DOI). This method got the job done but left very inadequate documentation. He was tasked with the installations in the Vickers Viscount mentioned above and following the crash and obtaining a replacement, David Henderson ordered the drawings to be reissued to the workshops. He was therefore horrified and extremely annoyed to learn that there were no drawings, just Norman's scanty DOIs.

Wedding

On 3rd August 1963, Carole and I were married at St Clements Church in Worcester with a reception at the Talbot Hotel, a coaching inn in the Barbourne district of Worcester. After an idyllic honeymoon in Cornwall, which was not entirely trouble-free including Carole getting seriously sunburnt and our Renault 750 breaking down on the return journey, we settled down in our new home at 12, Coxlea Close, Evesham.

The Demise of the Renault 750

After about 12 months in our new home, the Renault spectacularly failed its MoT to the extent that it was beyond economic repair. After six carless months, I purchased Mum's 1960 Triumph Herald for the trade in price she was offered in the deal for her new Triumph Herald.

This left the problem of disposal of the 'deceased' Renault. To achieve this, I removed the engine, transmission, suspension and wheels, and cut up the body into manageable chunks for transport to the local scrapyard. Taking on such a monumental task seems beyond belief now but I suppose paying someone else to do it was beyond my finances then. I have also always had a strong tendency to do everything myself. That is until rheumatic arthritis intervened a few years ago.

Tall Paul, Small Paul and Simon de Montfort

In the office, I had two colleagues called Paul. To distinguish them we called them Tall Paul and Small Paul. I became fairly close friends with the latter and he once asked me a favour. He was a bachelor and had a MG Midget sports car. This car was nearly identical to the Austin Healy Sprite Mk II and sceptics who disapproved of 'badge engineering' christened them 'Spridgets'. Anyway, Paul had arranged to have two aunties stay with him for a weekend and needed a four-seat car. As I then had the Triumph Herald, the solution we agreed was to swap cars for the weekend.

On the Sunday, we decided to take the Midget for a spin. This needed a bit of arm twisting on my part as Carole was a bit reluctant, being a nervous passenger and not keen on sports cars. After the Triumph Herald, the Midget had very hard, almost none existent suspension and very 'twitchy' handling and on the return journey we were approaching the 'Leicester Tower' road junction when the car spun off. Luckily there was a very wide flat verge and we came to rest with no harm done except for rendering Carole in a state of shock. This was near to the site of the Battle of Evesham of August 1265, where Simon de Montfort met his end at the hands of the army of Prince Edward who later became King Edward I.

I had another car adventure with Small Paul. At that time, the MoD held occasional auctions of retired military vehicles at a depot at Mottingham in South East London. These were notified in MoD official notices and Paul spotted a number of Mini Mokes listed. He decided he would try to purchase three; one to keep and two to sell at a profit. He roped in me and John Holt, another colleague in the Design Office, and we drove to the auction in the car of a friend. Paul succeeded in purchasing three Mini Mokes as planned and we drove in convoy back to Pershore.

The Suicide

Shortly after my wedding there was an ultimately sad series of events involving the staff of the design office. One of the two tracers in the design office was a girl called Judy, who was a pretty girl and knew it. This resulted in her being frequently chatted up by the younger male members of the office, not including me, it has to be said. I being somewhat wary of her, not being attracted to 'feme fatales'. In any case, Carole more than fulfilled my needs in that direction.

When a new recruit from the apprentice scheme called Roger arrived, he also immediately started chatting up Judy. He was more persistent than the others and this resulted in her ditching her long-term boyfriend for Roger. Very shortly after, we heard that the boyfriend had hanged himself in Tyldesley Wood in a ritualistic suicide. Not knowing any of the motives and emotions of those concerned I cannot make any judgement on this unhappy incident. Several years later, I met Roger again when he moved into the house next door to Carole's parents, with his wife, who wasn't Judy.

Snow and Fog

During my first winter at Pershore we suffered from severe snow storms just before Christmas, which caused the staff bus which ran from Malvern to Pershore via Worcester, to skid off the road down a bank, into a wood.

The airfield area, being located thirty miles south of the then heavily industrialised Birmingham/ Black Country/ Wolverhampton areas, was subject to very heavy fogs and 'smogs' in the 1950s and 60s. My journey home from work was disrupted by thick fog on two occasions.

On the first occasion, I was returning to my parents' home at Spetchley where I was living at the time, when I hit a bank of fog at about half distance. The Renault 750 car I owned at the time had very weak headlights and a six-volt electrical system which was probably an advantage in fog but I was still forced to drive very slowly and soon had a queue of cars behind me. I signalled for them to pass me but no one took up the offer. I recognised the steep hill on the approach to Spetchley and when I passed a field gate a few yards before our house I crossed over to the 'wrong side' of the road to locate the entrance to our drive and was followed by the queue. When I turned into the drive, the queue came to a halt and waited a while before proceeding gingerly down the road.

On the second occasion, I was returning home to our newly married home at Evesham. The route to Evesham involved crossing the traffic light controlled main runway and I became disoriented and lost crossing its wide open spaces. I eventually found the grass verge and followed it in search of the road. After quite a while, I found the road and was surprised how far down the runway I had strayed.

Solo whist and Archery

The normal lunch break activity in the design office was solo whist with a 'first team' and spectators. I was occasionally invited to play in the absence of a 'first team' member, but not too often as I was not a very accomplished player of solo whist.

There was an officer's mess just down the road from the main entrance and they offered us the use of their archery equipment. We quite enjoyed our archery sessions although I never really got the hang of it and was surprised at the strength required to pull the bowstring back to your ear.

Motor Racing

Two of my colleagues John Holt and David Gate shared my enthusiasm for motor racing and we attended meetings at Silverstone, Aintree and Oulton Park. Because we were all more interested in the technical aspects of the cars than the outcome of the race, we attended on practice days rather than race day. Also, the entry fee for practice days was a tiny fraction of that of race day, typically £1 for a car and all its occupants.

In the '60s, there was no structure to the practice and no separate qualifying sessions, the fastest laps achieved over the two days of practice deciding the starting grid positions. Also, the paddocks were not today's heavily guarded buildings alongside the pits but a field or orchard alongside the racetrack and entry to the paddock was free. You could examine the cars and chat to mechanics and the likes of Jim Clark, John Surtees and Jack Brabham.

On our only visit to Aintree, we were in the main grandstand watching the action when we strangely got into conversation with Mrs Topham who was the owner of Aintree, which was also the venue for the Grand National horse race. She admitted to not being particularly interested or fond of motor racing and quizzed us on our enthusiasm for the sport.

Career Move

With the cancellation of TSR2, there was a temporary lull in projects which caused me to start looking for ways to enhance my career. I therefore applied for four job advertisements in the engineering press and the civil service vacancy lists.

Two of these involved car journeys of over 100 miles and as this was in our carless period, our Renault 750 having failed its MoT and was beyond economical repair, I needed to hire a car. Fortunately, the foreman of the mechanical workshop at RRE Pershore ran a car hire business on the side with a fleet of four Volkswagen Beetles. These I found slightly odd to drive. I was used to the rear engine configuration with the Renault 750 but found the throbbing air-cooled engine and bottom hinged pedals difficult to get used to. I took Carole with me for both of these interviews.

The first job interview was at the Ford Motor Company test facility at Aveley in Essex and when we got there we were confronted with a flat featureless urban landscape. It was some time before the interview was due so I parked the car at the railway station which had a pedestrian bridge which was the highest point in the area. The view from there featured a sea of pre-fabs almost as far as the eye could see. Carole took one look at this and said, "I don't want to live here." I was inclined to agree and we decided there and then that I would probably not be taking up this job.

The job interview was accompanied by a tour of the test centre and this also proved to be a disappointment. They seemed to be testing mundane aspects of the car such as a test rig which opened and shut the car doors. This was a long way from my fondly dreamed notion of developing the Lotus Cortina alongside Colin Chapman.

The second interview was at the BRM racing car establishment at Bourne in Lincolnshire. The area around Bourne also turned out to be flat and featureless but this time rural. The BRM premises were grim grimy Victorian buildings and the job interview was also a disappointment. The job they offered was a gearbox designer, an area in which I had no experience, except for vaguely remembered lectures on epicyclical gears during the theory of machines subject of my HNC course. This did not seem to put them off eliciting the response that 'you'll soon get the hang of it'. We decided that this was not for us either.

The other two job interviews involved a rail journey from Evesham to Paddington and a London underground trip to the embankment station. They were both for graduate engineer posts, one with the Ministry of Defence (MoD) and the other with the Ministry of Supply (MoS). The interviews were at the Adelphi and Prospect House respectively. Both were large office blocks off John Adam Street, overlooking Victoria Embankment Gardens and the River Thames.

As a result, I was offered the MoS post, located at the Royal Aircraft Establishment (RAE) at Farnborough, which I accepted. This involved a job transfer within the MoS so generous financial assistance with removal expenses was offered. These included funding of three days in the Farnborough area to search for a house.

When we started searching at the estate agents in the area, it soon became obvious that houses in the area were far more expensive than at Evesham. We eventually found a house at Bordon, 15 miles south of Farnborough, which we could afford with financial assistance as part of the removal expenses.

Canberra WH 953 – I composed this picture from two photographs and my memory of the RRE experimental radar appendages. The airfield photograph was of an abandoned and somewhat derelict airfield, which is reflected in my drawing. So perhaps it is the ghost of Canberra WH 953 which flies above.

Jim Clark in the Lotus paddock alongside two Lotus 25s at the practice day for the Oulton Park Gold Cup of September 1962 which he won.

Part 3
Graduate Engineer

Chapter 6
Networks, Alarms and Concorde

Royal Aircraft Establishment – Farnborough

Graduate Engineer

Projects included:

1) Introduction of a network planning and optimisation system based on the US Program Evaluation and Review Technique (PERT) and the UK Critical Path Analysis (CPA) for test facility major overhauls.

2) Design of a centralised alarm and condition monitoring system for major plant items.

3) An Organisation and Method study of Aircraft Department for the RAE Chief Engineer.

4) Studies for a Concord thermal stress rig.

Education/Training included:

1) Part 3 of I Mech E – Industrial Administration.

2) Electronics for Mechanical Engineers at College of Electronics-Malvern.

3) Analogue Computers at Sheffield University.

4) Vacuum Systems at Edwards High Vacuum

Attained Associate Membership of Institution of Mechanical Engineers.

The RAE site at Farnborough

I duly reported for duty at the main office block at RAE Farnborough in mid-June 1965 and was introduced to Mr Johnson-Hall, Assistant Director, Workshops, who was to be my tutor. He allocated me to 36 Department, which was responsible for the maintenance and operation of the establishments major test facilities, such as wind tunnels.

I also enrolled in the Industrial Administration course at Farnborough Technical College which was part three of the Institution of Mechanical Engineers academic requirements.

After the scenic Malvern Hills location of RRE Malvern and the wide-open spaces of Pershore Airfield, the 'Factory Site' at RAE Farnborough had an air of 'dark satanic mills' about it. In my tour of the establishment facilities, the main impressions left on me were a mixture of the ancient, the weird and equipment commandeered from Germany at the end of the war. These included a bank of compressors with huge flywheels which by all appearances dated from when George Stevenson was a lad. I was also struck by a giant mercury arc rectifier with a permanent 'lightning strike' between electrodes reminiscent of a Frankenstein laboratory.

The first few weeks at Farnborough were also a somewhat unhappy period for me as I was separated from Carole while the sale of our Evesham house and purchase of our new home at 9, Hollybrook Park, Bordon progressed. During this period, I lived in a staff hostel, a mile or so north of the establishment, from Monday evening until Friday morning, returning to Evesham and Carole at weekends. The hostel was a gloomy wartime H block building, unlike the elegant Park View Hostel at Malvern and this added to my gloom and despondency.

The evenings tended to be rather boring as I had no desire to go to cinemas, pubs or restaurants alone. This left going for walks as my only recreation but the area was nearly all housing estates except for Mychett Common and Lake which had a mansion in the grounds which I seem to remember was associated with the French Emperor Napoleon III. Presumably, after he had been thrown out of France on the formation of yet another republic. He is buried in the nearby St Michaels Abbey. My knowledge of French history, and the language as well, come to that, is sketchy almost to the point of nonexistence.

I was also disappointed by the lack of interesting aircraft at Farnborough as most of the flight test activity had been transferred to RAE Bedford in 1955. Most of the flying at Farnborough was carried out by the Empire Test Pilots School which put selected RAF and RN pilots as well as a few foreign pilots through an intense twelve-month training programme to qualify them as test pilots. This organisation transferred back to its initial home, the Aircraft and Armament Experimental Establishment at Boscombe Down in Wiltshire in 1968.

The Lodger Units

During my period at Farnborough the following lodger units operated from the site;

The Air Accident Investigation Branch

The Army Aviation Squadron

The Army Personnel Research Establishment

The Empire Test Pilots School

The RAF Institute of Aviation Medicine

The Meteorological Research Flight

The Comet Tank

There were also a fair number of 'no longer in use' facilities scattered around the site including the very large water tank used to fatigue test a comet fuselage after the air crashes of the 1950s. The engineer who led the team that built this tank had become something of a 'folk hero' to the supervisors and craftsmen of 36 Department who regaled me with stories of his exploits. He came across as a hard driver who went to extraordinary lengths to keep the project to its timescale. He was also inclined to award the team for long overtime hours by importing crates of beer for their consumption.

Nevertheless, there was quite a lot of 'up to the minute' work going on at the time, including the development of carbon fibre structures, an early 'fly by wire' system fitted to a modified Hawker Hunter aircraft and the ogival delta wing for the Concorde. A new structural test facility in support of the Concorde programme was nearing completion at the west end of the establishment.

Some RAE Characters

As well as the above-mentioned engineer, I was told about a couple of other notable characters. The first was Miss Tilley Shilling, a name she retained in spite of being married. She had gained notability during the war by solving a carburation problem on the Merlin engine, which powered the Spitfire and Hurricane, which caused cut outs during combat manoeuvres which involved sustained negative G. Her solution involved the insertion of a metal disc in the fuel system, which became known as 'Tilley's shilling'. She was small in stature but a formidable, domineering woman who did not suffer fools and liked to ride fast motorbikes.

The other was a senior scientist who suffered a major nervous breakdown and was sectioned to a local mental hospital. On his discharge, he was given some sort of release document which he had framed and hung up in his office. He boasted that he was the only person in RAE who had been certified sane. I understand that his desk was decorated with a very large machete style knife so a visit to his office must have been a bit intimidating.

We Become a Family

On 18 March 1966, Carole gave birth to a boy, James Michael at St Luke's Hospital in Guildford. Carole took motherhood in her stride, like everything else that life presented her with, in spite of a difficult birth which involved an emergency ambulance drive from Hazelmere Maternity Hospital to St Luke's Hospital in Guildford and later, rather alarming setbacks like projectile vomiting.

Institution of Mechanical Engineers (I Mech E)

The graduate engineer training was geared to obtaining associate membership of an engineering institution, in my case the Institution of Mechanical Engineers (I Mech E). The RAE had a quite active branch of I Mech E which organised factory visits and lectures at the staff canteen. The former included visits to Johnsons Wax and a local printing works. The most interesting lecture I remember was given by a German engineer called Dr Ansdale who did a lot of the development work on the Wankel engine.

I also fondly remember a dinner dance at a Bournemouth seafront hotel organised by a south of England young engineers' branch. It included an overnight stay and I attended with Carole and baby James. The hotel provided a baby watching service while Carole and I wined, dined and danced the night away. The next day before returning home we walked along the beach pushing James in his pram in the freezing cold of a late November morning.

The 750 Motor Club

At this time, I joined the 750 Motor Club which had an active branch in south east Hampshire. They had monthly meetings at a pub near Butser Hill, which sometimes featured talks by notable people from the world of motor racing who had started out competing in club events. They also held trial events in the locality and we attended both on a regular basis for quite a time. They had a quarterly magazine which usually had very interesting technical articles in it.

The main reason for the club's existence was to run circuit racing and trials events for homebuilt cars featuring Austin 7 engines, later to be joined by Ford Ten 100E engine cars. At about this time, a switch was made to the Ford 105E pushrod engine and the engine of the current Reliant 3-wheeler.

Quit a few famous names in the motor sport world started out competing in 750 Motor Club events most notably Colin Chapman of Lotus Cars, whose first car, the Lotus Mk 1 was a trials car using an Austin 7 engine and modified chassis.

The club's main activity was circuit racing to their 1172 formula. Two cars racing at the time in this formula caught my attention. The first was the U2 designed and built by Arthur Mallock. It was a very crude device, somewhat resembling a children's soapbox racer with a very basic technical specification and apparently questionable build quality. However, it was amazingly fast and wiped the floor with more sophisticated, modern opponents. The other one was a car called the WEV which was an acronym for 'worm's eye view' as the owner had gone to great lengths to make the car as low as possible.

The BMW 700, Mini Cooper and 'Rosemary Smith'

At about this time, we made the first of several trips to London. The trip was to North London to exchange the Triumph Herald for a BMW 700 Sport Coupe at a car dealer located on the North Circular Road near Chingford. The journey consisted of the A3 to Surbiton, then to Kingston and Richmond, joining the A406 North Circular Road and following it to our destination. No M3/M25 in those days.

The BMW 700 was a very attractive and technically interesting coupé with a BMW motorcycle engine at the back and had a lively performance. It was however a bit impractical for a couple with a baby, especially as James' pram was a large coach-built pram which I had to totally dismantle to get it in the car. I did not care about this inconvenience and we managed. It remains one of my all-time favourite cars, unlike modern BMWs which are conservative and a bit boring, with the exception of the electric cars. However, the car was not without its problems.

The first one was a slipping clutch, which involved the purchase of a replacement clutch and the removal of the engine to gain access to the clutch. To do this I had jacked up the rear of the car with it located in the drive and undid the engine mounts allowing it to be lowered. Having removed the bolts attaching the engine to the tans-axle, I was lying on my back under the car pulling the engine back to release it from the gearbox input shaft when it suddenly fell on my chest causing acute pain. I was lucky not to sustain broken ribs.

While all this was going on, Carole, who was taking James for a walk in his pram, returned just in time for me to let out a cry of pain and start to shuffle out from under the car on my back with the engine, which was a lot heavier than I anticipated, on my chest. Carole let out her own little cry as she dashed towards me. I can't remember what she said next, but it was probably something like, "Oh, Mike, what have you done now, you silly bugger."

Having completed the clutch replacement, the car ran OK for about 12 months before disaster struck in the form of a loud knocking from the engine. I consulted BMW Concessionaires at Portslade near Brighton and they said it was a known problem of big end failure, caused by the conversion from the dry sump lubrication of the motorcycle engine to wet sump for car application and the use of a centrifugal oil filter, which over a period of time filled up with sludge and stopped filtering. The outcome was a drive down to Portslade where BMW had to replace the entire roller bearing crank assembly for the princely sum of £75 which was a lot of money in the 60s. Their suggested solution to the problem was to use Shell detergent engine oil which was designed for tractor engines and could only be purchased in five-gallon drums and change the oil at half the recommended period. By then, I had enough and exchanged the BMW for a Mini Cooper.

As well as mechanical problems, we suffered accidental damage on one occasion. We were waiting behind another car to turn onto a main road. When the car moved off, I followed but he suddenly stopped and I went into the back of him sustaining a dent in the front of the car. The other driver and I examined the damage, which was slight on his car, so we agreed not to involve Insurance.

Fortunately, the husband of one of Carole's friends was a panel beater who had been trained at Tickford, the coachbuilders, who made special bodies for Rolls Royce. He did a splendid job on the repair but was colour blind, so I had to do the colour matching when we blended the paints.

I later traded the Mini Cooper, which was a bit of a disappointment, for a Singer Chamois. This was a posh version of the Hillman Imp. I named the car 'Rosemary Smith'. The lady concerned was a rally driver from Ulster, who achieved considerable success in international rallies driving a works Imp. I'm sure I was not the only person who named their Imp thus and there were probably quite a lot of Imps called 'Rosemary Smith' in the 60s.

The Epic Quest for A Sports Car

About mid-point in our stay at Bordon I had a yearning for a sports car and started looking for one in the car pages of Exchange and Mart, which was the eBay of those days. The first advert which interested me was a Lotus 11, located at an address in Frome. We called in on Carole's gran and she gave us directions. The address was in the Keyford area of Frome and turned out to be a small holding with a steep driveway with high banks leading to the house and barns. We were shown into one of the barns and the Lotus turned out to be a pile of rusty and battered bits. Carole's first reaction was to burst out laughing. This was quickly followed by a straight face and 'We're not having that'. I did not need any persuading however and we made our excuses and beat a hasty retreat.

Our second stab concerned a Marcos Gullwing located at Esher in Surrey, whose main claim to fame then was that George Harrison had a house there. When we arrived, the high street seemed to consist almost entirely of antique shops and hairdressers. We located the house which was a mini mansion and this time the car was at least roadworthy. It was a bit tatty however and there were a number of holes cut in the internal panelling for no good reason that I could discern. It was also a bit pricey so we declined on that one.

I next came upon an article and advert in Autosport magazine about the recently introduced Mini Marcos. It was available as a body/chassis kit to which the customer fitted separately sourced mini components. I decided to visit the Marcos factory at Greenland Mills, Bradford on Avon to view a car. I believe Greenland Mills is now luxury apartments. When we arrived, there was no sign of life and no cars to be seen.

A few months later I spotted an advert in Exchange and Mart for a Mini Marcos by a Flight Sergeant at RAF Upavon. We decided to combine a viewing with a visit to Carole's relations at Warminster. When we arrived at the RAF station guardroom, I showed the duty sergeant the advert and he despatched someone to find the flight sergeant. However, there was no sign of him or the Mini Marcos on the camp. I left my address with the duty sergeant but never heard from the flight sergeant so presumably he had already sold the car.

Carole was very patient with all this and we usually combined the trips with something she would enjoy like a visit to her relatives or a beauty spot. Also, she was probably intrigued to see what I would come up with next after the fiasco of the Lotus Eleven.

I eventually purchased a Berkley sports car. This was a tiny sports car powered by a 350cc two stroke motorcycle engine, made by a caravan company at Biggleswade in Bedfordshire. Its body/chassis was a mixture of fibreglass, aluminium and steel and was a light as a feather. One person could pick up the back of the car and wheel it about like a wheelbarrow. This was just as well as the car did not have a reverse gear, due to its motorcycle gearbox.

There proved to be two other drawbacks with the car. Firstly, having aluminium and steel components bolted together with no corrosion preventative treatment gave rise to electrolytic corrosion. Secondly it was the noisiest car I have ever owned with an exhaust note like a machine gun which became tiresome. Carole was relieved that I had bought a cheap and inoffensive sports car rather than a raging monster but otherwise took no interest in it and refused to go in it. After about six months, I sold it to a man who collected it in a Morris Minor with a roof rack. We lifted the car onto the roof rack, tied it down and off he went.

C.P.A/PERT

My first task in 36 Department was to provide a plan for a major overhaul of the Hypersonic Wind Tunnel using the fairly new technique of Critical Path Analysis (CPA), also called Programme Evaluation and Review Technique (PERT) in the USA. I had only a vague knowledge of this management technique so the first thing I did was go to the establishment's very extensive library to read up on it. The basis of CPA/PERT is to break the project down into its individual tasks and represent them as a logical network. The estimated times for each task are then added and the longest path through the network is deemed the 'critical path' and defines the project timescale. It is then essential that adequate resources are allocated to the tasks on the critical path to ensure that the timescale is met. Also, if it is desired to shorten the timescale, more resources can be applied to tasks on the critical path, which may result in another path becoming the critical path and so on.

The Hypersonic Wind Tunnel was designed to test small models at speeds of the region of Mach 5 and consisted of compressors, a heat exchanger and a fast-acting valve leading to the working section with vacuum chambers at the other end. The fast-acting valve consisted of a plate with a machined cross at the centre. The thickness of metal at the base of the cross was calibrated for the valve to break at a given pressure differential giving the desired speed of airflow at the working section. The desired airflow only lasted about a second and was recorded using a technique called schlieren photography.

My first task was to break the project down into individual tasks. To do this I needed the information from the team supervisor and his charge hands. In spite of being very sceptical about CPA/PERT, they were extremely helpful in providing this information. Having compiled the network plan, I consulted the team on its logic and explained the critical path and timescale to them.

The project then proceeded and was keeping to the plan when disaster struck in the form of a heat exchanger chamber which had collapsed under the pressure differential applied during operation. The material of this chamber was a nimonic alloy which had a very long lead time, causing the project to be halted for several months.

However, the project plan was deemed to have worked quite well and the next project plan was entrusted to another graduate engineer to plan using a computer. Meanwhile, I had been moved on to another department and another project.

Shot Blasting, The Electron Beam Welder and NC Machines

When the Hypersonic Wind Tunnel overhaul came to a grinding halt, I was asked to look into the design of a chamber for shot blasting rolled steel Joyce (RSJ) sections and assist in the commissioning of a newly acquired electron beam welder. This was very much the latest technology and ours had been supplied by a firm from the USA called Skiaky. Unfortunately, the equipment was plagued with teething troubles and was obviously underdeveloped by the firm with ongoing serious problems which were still unresolved when I moved on to another department.

While I was in Workshops Department, I spent some time in the machine shop to see the introduction of numerical control technology to some of their milling machines. These were fairly crude early devices with an electronic computer next to the machine tool in a cabinet as large as the machine tool itself, which was fed with a programme on punched paper tape.

Progress Meetings

A regular feature of my period in Workshops Department was progress meetings. I found these to be very tedious and this feeling remained with me for the rest of my career even when much later I was chairing meetings. I sat through these meetings which usually lasted all afternoon with a mixture of frustration and boredom. On one of these meetings to relieve my boredom, I was balancing myself on the back legs of my chair when I overbalanced, fell over backwards and hit the floor with a loud crash, waking everybody up. Surprisingly, the chairman, Vic Gurney did not show any annoyance but treated it as part of the proceedings. This was especially surprising as he struck me as a somewhat pompous individual, a bit like Captain Mainwaring of Dad's Army. I imagined him muttering 'stupid boy' under his breath. The other thing I remember about him was his immaculate vintage Riley car.

Centralised Alarm System

My next department was called Heat, Light and Power (HLP) and they did 'exactly what it says on the tin'. The project I was given was a centralised system of alarms for the establishments boiler houses, electrical sub stations and major plant facilities. I assume that one of the drivers for this requirement was the very heavy electrical power consumption of some of the establishments' major plant facilities. The running of these was controlled by the nearby National Gas Turbine Establishment (NGTE) which placed an even heavier demand on the local grid. First of all, I compiled an inventory of these alarm systems and what they consisted off. I then consulted the various managers in the department on which of the alarms needed to be centralised and the best location for this alarm station. Having plotted these on a map of the establishment we needed to decide on the best way of transmitting the alarms to the central station. The obvious solution was to use telephone land lines and this was agreed as the way forward.

Thermal Stress Facility

I was then transferred to Structures Department and the new structural test facility built in support of the Concorde programme.

I was not given a specific project but was attached to a team who were looking at the thermal stresses likely to be encountered during the likely flight envelope of Concorde and the design and construction of suitable test rigs to investigate this phenomenon.

At the time of my graduate engineer training at Farnborough, the new Harold Wilson Labour government made sweeping changes to the organisation of the civil service including a new Ministry of Technology under Antony Wedgewood Benn, to which the RAE became part. Antony Wedgewood Benn proved to be a very good minister for us and encouraged the research and development establishments like RAE to get involved in projects to help UK industry. One of these was the investigation of a serious vibration problem on the Hillman Imp car using the new structural test facility. Also, the 24-foot wind tunnel was used to test large scale models of tall buildings, cooling towers and even the effectiveness of trees as wind breaks.

BAE Filton

Our work in support of the Concorde programme involved several visits to BAC Filton for formal meetings. The journey down to Filton usually started out early and included a stop for breakfast at the Pretty Polly restraint in the high street at Marlborough. I found the meetings to be long winded, tedious and boring, but this was compensated by very pleasant lunches. The canteen building at Filton was a structured by class affair with the workers' canteen on the ground floor and several staff canteens on upper floors with the senior staff mess on the top floor. It was the latter which we as visitors were entitled to but the BAE staff we visited, had a ploy of not booking us in until the last minute when the mess was likely to be fully booked. When this happened, we instead had a more 'liquid' lunch at the Lamb and Flag at Patchway, which was a rather splendid hostelry which did very good meals.

Short Courses

During my graduate training I was sent on several short courses. The first was an analogue computer course at Sheffield University. At the time, these were mainly used to simulate dynamic situations but its use was being rapidly taken over by digital computer technology. At that time, Sheffield was famous for manufacturing a multitude of items in steel, so I decided to buy Carole something made in steel from Sheffield as a gift. I found a shop in the town centre which specialised in this sort of stuff and purchased a meat carving set comprising a carving knife and fork with a stainless-steel platter with spikes to retain the joint being carved. Carole was quite pleased with this gift but on removing the packaging spotted the legend 'made in Japan'.

The most pleasant course for me was Electronics for Mechanical Engineers at the College of Electronics at Malvern. This proved to be very interesting with the new developments in semiconductor technology taking over from thermionic valve technology. For this three-week course, I stayed at Carole's parents' house in Worcester with Carole and James and we enjoyed our evenings and weekends revisiting family, old acquaintances and local attractions.

The vacuum technology course at Edwards High Vacuum in Crawley was the most interesting one for me as the technology involved was new to me. This was during my period in 36 Department and a lot of the new test facilities at RAE were using vacuum pumps and vacuum measuring equipment, usually supplied by Edwards High Vacuum. The company's vast range of products ranged from simple mechanical pumps to huge diffusion pumps.

Graduate Review Board

At this stage, I was invited to attend a formal interview board to review my progress. The outcome of this was a recommendation that I be given a specific demanding project, which, on completion would be followed up by a formal promotion to junior engineer interview board.

One of my fellow graduate engineers going through the same procedure was asked at his interview for his views on the UK joining the Common Market. His response to this was 'You know what they say about this, Wogs start at Calais'. He intended this to be a satirical joke but it misfired and was not well received by the board. Needless to say, he failed the interview, but this statement perhaps reflects the attitudes of today's Brexiteers.

This same individual had an obsession about Freemasons. He was of the opinion that there were a lot of them in senior management posts at Farnborough. I do not know if this was true or not and was not aware of any. He was of the opinion that to get a top job at Farnborough you had to either be a Freemason or Welsh and there were indeed quite a few Welshmen in top jobs at Farnborough at that time.

Experimental Aircraft Servicing Department

As a result of the recommendations of my review board I was invited to an interview with the establishment's chief engineer, Mr 'Bertie' Beer. The outcome of this was that I was tasked to carry out an 'organisation and method' study of Experimental Aircraft Servicing Department.

The first thing I did in response to this was request an interview with the head of the department and he responded by providing me with an organisation chart and calling in his engineers and supervisors to introduce me and ask them to provide every co-operation.

In my study of the department, I found that they were only responsible for the servicing and preparation for flight of the establishment's aircraft, including those of the Empire Test Pilots School and the Meteorological Research Flight. Aircraft modifications and installations were designed and manufactured by other departments design offices and workshops.

My report to the chief engineer, included the recommendations that the department should have its own design office and workshops specialising in aircraft structures. To be honest, my recommendations were heavily influenced by what I knew of the organisation at RRE Pershore.

My report was accepted in full by the chief engineer and the department and the reorganisation set in motion almost immediately. I was then successful in my promotion review board.

Promotion to Junior Engineer

I then scanned the junior engineer vacancies at RAE and chose the one at the Aircraft Escape Systems Section of Engineering Physics Department. At an interview by the section leader, I was accepted for the post and started the following week.

The BMW 700
1967

Carole with the BMW 700

Part 4
Junior Engineer

Chapter 7
Ejection Seats

1968–1971
Royal Aircraft Establishment – Farnborough

Engineer–Escape Systems Section–Engineering Physics Department
The design of test vehicles and facilities for the development of aircraft escape systems and the planning, organisation, implementation and reporting of trials.

Projects included:

1. Stabilisation of ejection seats using thrust vectored rockets controlled by fluidic sensors.
2. Stabilisation of ejection seats using a guide surface parachute.
3. Trials at Pendine Ranges of a rear crew escape system for V Bombers
4. Trials at Pendine Ranges of a German Air Force F104 escape system.

Education/training included;
RAF Introduction to ejection seats at RAF Sealand.

RAF Sealand and Ejection Seats

As soon as I started my new job, I was sent on an ejection seat training course at RAF Sealand, near Chester as my knowledge of ejection seats was somewhat limited. I found the subject and the technology involved fascinating as it was all new to me. The following description applies to the Martin Baker ejection seats which were current with the RAF at the time (late 1960s).

Martin Baker seats were fitted to all the RAF and RN fast jets with the exception of the Folland Gnat advanced trainer which was fitted with Folland's own seats. These were developed from a Saab design which was in turn developed from a late WW2 German design. Martin Baker seats were also fitted to most western fast jets with the exception of Saab and Lockheed aircraft which were fitted with their own seats.

The means of ejecting the aircrew was the ejection gun which was powered by cartridges loaded with nitro cellulose propellant. The main cartridge at the top of the gun was initiated by a hammer mechanism released by a wedge pulled out by a cable connected to a face blind handle in the seat head box, above the pilot's head. The face blind incorporated in the firing handle was to protect the pilots face from the blast of ejection and entering the airstream at speeds of up to 600 knots. There was also an alternative firing handle at the front of the seat pan. There were two pairs of secondary cartridges at the side of the gun which were initiated by exposure to the heat and pressure as the telescopic gun deployed.

Before the ejection gun fired the aircraft, a canopy was opened and despatched by a pair of telescopic guns followed by a fraction of a second delay. During deployment the seat was guided by the seat rails on either side of the ejection gun and at separation a drogue gun on the side of the head box fired a slug which deployed a stabilising parachute which stabilised and slowed the seat and occupant down to a safe speed for deployment of the main parachute. The main parachute was housed in the backrest of the seat and was deployed by the stabiliser parachute at the same time as the pilot was separated from the seat.

The whole of this sequence took about two seconds and the pilot was subject to 20 g during the gun stroke. This resulted in damaged vertebrae in a high percentage of ejections. The sequencing and timing of all these operations was controlled by a clockwork timing box and all these systems were self-contained on the ejection seat and independent of the aircraft systems.

Testing Ejection Seats

The Escape Systems Section had two facilities at its disposal.

The Ejection Seat Test Rig was the same as the Martin Baker rig but with an extended ramp. It was used mainly for testing personal survival packs (PSPs) during the period of my employment in the Escape Systems Section. PSPs were located in the seat pan of the ejection seat and doubled as a seat cushion. The reason for these tests were to check that the 'springiness' of the PSP did not add to the G force on ejection. Two types of seat cushion were tested in this period. The first was a sheep pelt which was very popular as an accessory car seat cover at the time. The second was a bean bag cushion.

The long test track at the Proof and Experimental Establishment (PEE) at Pendine in South Wales consisted of about a mile of narrow-gauge railway mounted on a continuous concrete plinth. Ejection Seats were tested by firing them from a rocket powered sled and monitoring their trajectory with kinetheodolites and cameras.

Two types of sled were employed, both of RAE design. A recoverable sled was used for ejection speeds of up to 600kts. This was a rugged design consisting of a large section square steel tube frame clad in aluminium alloy panels. It had provision for boost rockets at the back and retro rockets at the front. For speeds above 600kts, an expendable sled was used which simply crashed off the end of the track after the ejection. This was a lightweight sleek projectile with a vertical chisel front manufactured from thin plywood clad honeycomb panels by Marshalls of Cambridge.

The boost and braking for the sleds were from a selection of solid fuel rockets of various thrusts and durations, supplied by the PEE Range. The rockets were selected to provide acceleration to the desired speed, maintaining that speed for two seconds and bringing the sled to a stop before the end of the track using calculations developed by the Escape Systems Section.

Instrumented test dummies were used for tests on both the Test Rig and the long test track. The RAE used dummies of their own design which was a versatile dummy with separate main body parts which could be dismantled and separately ballasted to provide the desired body weight. I was always surprised at how heavy these body parts were. Martin Baker and A&AEE used USA supplied Alderson dummies which looked more realistic but were not as versatile.

The aircraft escape system trials fitted into three categories.

1. Those in support of the sections R&D programmes.

2. Those in support of A&AEE Boscombe Down release to service of new aircraft/ejection seat combinations.

3. Support for Martin Baker programmes.

Engineering Physics Department

The department started life as Structures and Mechanical Engineering Department but separated as Mechanical Engineering Department in the 1960s although the separated departments shared a design office and some other functions. During the late 1960s, it changed its name to Engineering Physics Department for reasons which were never explained to me. Perhaps, the new name sounded more scientific and scholarly.

The head of the department was an ageing academic individual called Templeton and my only memories of him are smoking a curly pipe and a very small Chihuahua dog; more of which later. There were four group leaders under him and the one responsible for the Escape Systems Section was a reserved taciturn individual called Geoff Spurr, whose hobby was white water canoeing which seemed somewhat at odds with his character. At the other end of the scale was a retired senior RAF officer, I think he had been an air vice marshal, who came in several days per week doing odd jobs. I guess he needed the company and something to keep him occupied.

The Escape Systems Section

My memory of the names of all the people concerned here is incomplete after 50 years elapsed time and where my memory has failed, I have simply not included them and instead used the remembered surname or Christian name or a job title.

The section leader was Desmond Rylands, a senior scientific officer who everyone called 'Des'. He was a likable but somewhat indecisive individual who lived at Woking and had two daughters.

He relied on an experimental officer, George Goodridge who had been in the section forever and had a vast, if somewhat superficial knowledge of ejection seats, associated equipment and it's testing. He also knew everyone there was to know in the aircraft escape systems field. He was a sort of undesignated deputy to Des Rylands and a likable Welshman.

There was another experimental officer called Dave Nosworthy, who I found a rather aloof individual and I never achieved any sort of rapport with him. He tended to pick and choose the jobs he was interested in and at the time I joined the section he had teamed up with the sections attached royal navy officer, had done a RN diving course and was taking part in underwater ejection trials.

The attached RN officer was Lt Cdr Jack Gaydon who was an engineering officer and a 'Clearance Diver' which was the navy's top-rated diver. He was mainly involved in underwater ejection trials in a large test water tank at Glen Fruin near Helensborough on the River Clyde. About a year after I joined the section he was transferred to an 'office job' in London and in disgust, took early retirement to start a diving school on the Scilly Ille's. He was replaced by Lt Eddie Day who I became great friends with and he and his wife Molly exchanged hosting several dinner parties with me and Carole.

There were a couple of scientific assistants. The first was Betty whose main job was analysing ejection trajectories from the films using established formulae. She was very welcoming to me and gave me lots of information about interesting places to visit. She was particularly enthusiastic about the seaside resort of Mudeford near Christchurch and much later in life Mudeford Sandbanks and Hengistbury Head became a regular haunt for visits to the seaside with our grandchildren. The other, Graham, was a likable enthusiastic young lad who willingly helped out on any job going and had an enthusiasm for the children's cartoon series Magic Roundabout.

There was an instrumentation technician, Terry, whose main responsibility was the instrumentation in the dummies which mainly consisted of three-axis accelerometers. The section was completed by a fitter who assembled the dummies and prepared the seats and sleds for trials. Unfortunately, he was showing early signs of dementia and everything he did had to be very carefully checked. He was later replaced by Ron, a cockney ex-London transport fitter who was very knowledgeable on diesel engines and bus transmissions but had no appreciation of aircraft engineering practice, so he too initially had to be very carefully monitored.

My first impression of the section was that there was a lack of direction with individuals 'doing their own thing' and spur of the moment decisions. Perhaps this was because this was my first experience of working in an experimental environment and I was used to working in a more structured ordered environment.

The section received essential support from two areas within RAE. The first was Structures design office who had a senior designer called Alf Rivers who I immediately gained a close rapport with and we worked closely in developing the Folland seats and the sleds.

The other was Photographic Services who had a photographer who was exceptionally good at following the seat trajectories. He was a quirky cockney individual who had a great sense of humour and was not beyond doctoring some of his films to provide amusing endings. As a hobby he was restoring a classic sailing boat and had an ambition to sail around the world in it. I wonder if he achieved this.

During the second school holiday of my period in the section we were joined by a work experience student, Mary Mullins, who was the daughter of the head of Structures Department. The latter was a very gifted academic who among his other attainments could speak numerous languages, both European and oriental. No one knew what to do with Mary, so she attached herself to me, which in turn attracted the young scientific assistant, so I ended up with two unasked for assistants. I initially set her to work analysing films supervised by Betty and she proved to be very quick on the uptake and a very useful assistant.

Pendine Ranges

The long test track was located in Pendine Ranges, which at that time was bestowed with the title of Proof and Experimental Establishment (PEE). It was located on the South Wales coast, halfway between Pendine and Laugharne. It was called the long test track, presumably because it was the longest test track on the site, but was in fact quite short compared with test tracks in the USA and the later Martin Baker test track in Northern Ireland. Hence the need for retro rockets to recover the sled instead of the more usual water trough braking systems.

The ranges were located between the A4066 road and the sea at a location called Llanmiloe and consisted of sand dunes on the south side and flat marshes on the north side. A feature of the latter during the summer was large quantities of yellow blooming wild irises. The long test track was located on the border between these habitats. There were offices, an officer's mess and a sick bay on the north side of the road under steep wooded hills.

A&AEE Release to Service Trials

These tended to be a bit of a circus with teams from RAE, A&AEE, Martin Baker and ML Aviation in attendance. RAE provided the sleds, photographic coverage and specified the sled rocket requirements, A&AEE provided overall co-ordination, instrumented dummies and kinetheodolites, Martin Baker provided the seats and brought along their own photographer, ML Aviation brought along the aircrew equipment and Pendine ranges provided the solid fuel rockets. These trials usually consisted of three runs, one at 200kts, one at 400kts and one at 600kts, all using the RAE recoverable sled.

The A&AEE team was led by Huw Thomas, a loud opinionated individual who was heartily disliked by the RAE, Martin Baker and ML Aviation teams. His boss, Don Buchanan, on the other hand was a very pleasant and reasonable individual. The ML Aviation team consisted of just two people, Morris Molloy and Nigel Hibbs who were both very likable individuals who were popular with everyone. The Martin Baker team remained something of a mystery to me as they tended to keep to themselves. There was one individual however, who was a real 'jack of all trades', at different times acting as technician, photographer, driver and even a pilot.

During my three years in the section there were three release to service trials. The first was the RN Phantom seats which were Martin Baker Mk seven seats. These were Mk five seats, updated by the addition of a rocket pack under the seat pan. This gave the seat what was called zero-zero capability, namely the ability to save the pilot at zero speed and zero altitude. Another advantage bestowed by the rocket pack was that the gun charge could be reduced, reducing the load on the pilot's spine from 20g to 15g.

The second was the Jaguar, an Anglo-French collaboration produced in both single and two seat forms. The seats used were the newly introduced Mk nine seats, which were developed from the Mk eight seats designed for the recently cancelled TSR2, but a simpler lighter seat. It struck me at the time as a very elegant design with the usual rugged but beautifully precise components. It dispensed with the head box face blind firing handle and now used just the previously secondary seat pan handle. The seat pan itself was strengthened to cater for the thrust from the rocket motor. The rocket motor, which was a multi-tube device filled with 1" diameter propellant tubes, was hinge mounted to rotate as the seat was raised and lowered so that the efflux projected through the centre of gravity of the pilot/seat assembly, to cater for different sized pilots. It burned for 0.25 seconds with a thrust of 6000lb.

The third was the Harrier vertical take-off and landing aircraft-ft which was fitted with a Mk nine seat fitted with a larger 2" propellant rocket pack, which burned for 0.5 seconds with a thrust of 6000lb. This seat was designed to save a pilot ejecting at a sink rate of 100 feet per second at zero altitude.

All these trials consisted of three test ejections at respectively 200, 400 and 600kts and all of the above were successfully completed, with no problems.

The Harrier Platform

With the introduction of the Harrier Mk nine seat with its 100ft/sec sink rate capability, it was felt that there was a need to test this capability. We were asked to look at this and came up with the idea of using a supply drop platform. To this end we approached the Supply Dropping Section who carried out supply drop trials using a Blackburn Beverly aircraft and came up with a plan to use one of their platforms to launch the Harrier seat from.

The plan was to launch the platform from the Beverly and use selected stabilising parachutes to achieve a sink rate of 100ft/sec, launch the seat and then recover the platform with its main parachutes. An initial trial was arranged with just the platform and its parachutes before carrying out an ejection trial. On the due day, the Beverly was flown over Larkhill Ranges on Salisbury Plain and the platform launched. It achieved the sink rate of 100ft/sec but the main recovery parachutes failed to deploy leaving the platform to plummet to the ground, leaving a nice bomb crater on the range. I do not know if the programme was pursued beyond this setback as by then I had moved on. In any case, I believe that Martin Baker came up with an alternative test method which was actually used to test Harrier seats.

The Multi-Role Combat Aircraft (MRCA)

At this time, the Multi Role Combat Aircraft (MRCA), which later became the Tornado was being jointly developed by the UK, Germany and Italy and we were invited to attend a meeting at BAE Warton to discuss the escape system. I was nominated to attend this meeting and my main memory is not of the meeting but the nightmare drive to and from the meeting up the M5/M6 in torrential rain in my Triumph GT6 with visibility being virtually non-existent.

The meeting itself cantered on the ejection of the canopy which was large and heavy. Martin Baker proposed using rocket motors, which some members of the meeting felt could endanger the aircrew. I think that Martin Baker would eventually allay these fears with a series of tests. In the longer-term, miniature detonating cord (MDC) was used to shatter the canopy and this method was employed on all future aircraft.

Trials in Support of Martin Baker

While I was in the section, we were asked to provide support to Martin Baker on two occasions. I'm not sure whether the first one was from Martin Baker, the German Air Force (Luftwaffe) or a joint one from both. It was to carry out the usual 200, 400 and 600kt ejections from a representative cockpit sled at Pendine.

The Luftwaffe had lost a number of Lockheed F104 Starfighters and their aircrew in air accidents, causing the F104 to be labelled 'the widow maker' by the Germans. Just why this happened was something of a mystery as the F104 had a good safety record with the US air force. The fact that the Lockheed ejection seat had failed to save the aircrew on some of these occasions made the Luftwaffe want to substitute the Lockheed seats for Martin Baker seats.

The Luftwaffe supplied a F104 cockpit which Alf Rivers and his team converted into a sled and designed a pusher sled to attach to the cockpit sled with retro rockets angled out so that the efflux was alongside the cockpit sled. This was quite a challenging project for Alf and his design team.

All three ejections were successful but when the retro rockets fired after the 600kt ejection the sled broke free from the test track rail, became airborne and cartwheeled down the range with the retro rockets still burning, the whole thing like a giant Catherine wheel. This was extremely spectacular and resulted in one of the German observers, an elderly individual, who had placed himself nearer to the test track than he should have been, having a heart attack and having to be rushed to hospital.

Sir James Martin

The second approach was from Sir James Martin himself and was in support of one of his attempts to persuade the RAF to fit ejection seats for V Bomber rear crew members. A meeting was arranged at Martin Baker to plan this and Des Rylands, George Goodridge and I attended.

When we arrived, we were ushered into Sir James' very large office and the meeting included a lavish lunch around a large table. There was very little technical discussion about the proposed trial and the meeting consisted mainly of Sir James waxing lyrical on all sorts of unrelated topics. I have already mentioned my dislike of meetings, considering them to be a very inefficient way of doing business, but on this occasion, I was actually entertained by Sir James rambling monologue and enjoyed the proceedings.

Sir James, who was known to people in the trade as Jimmy Martin but probably not to his face, was a larger than life character and very passionate about the role of his product in saving the lives of aircrew. At the time of this meeting, he was 77 years old and a bit deaf, but still played an active part in the design of his company's products. His office was littered with bits of ejection seats and a large radio to which he listened to news broadcasts at deafening volume.

After he deemed the meeting to be finished, he invited us to view his latest seats under development and on the way to the workshops we passed a young female employee who was somewhat scantily dressed for what was quite a cold day and Sir James immediately started to harangue her to 'Get some clothes on or you will catch a death of cold'.

The rear crew members of V Bombers sat in rear facing seats so we had to provide for this in our sleds and this was done by the simple expedient of turning the sled round. The trial was duly carried out and was very successful. However, Sir James' offer was still not taken up by the RAF.

Research and Development

During my period in the section I was involved in two of these programmes, both directed to stabilising the seat on entry into the airstream and separation from the ejection gun. The first was given to me to run in partnership with GQ Parachutes and involved the use of a guide surface parachute to stabilise the seat.

The second was the use of the then emerging technology of fluidics to thrust vector the rocket pack and this was given to Dave Nosworthy to run.

The Folland Ejection Seat

Our R&D programmes received a boost when twenty or so Folland ejection seats became available from Folland Aviation at Hamble, on the Solent. Why so many 'surplus to requirement' seats arose was not disclosed to me, perhaps a cancelled aircraft order. The Folland seat was similar in principal and operation to the Martin Baker product, but was flimsier in construction and a lot of the functions were operated by bicycle style Bowden cable rather than mechanical linkages which I did not like much. Also, the ejection gun consisted of a combustion chamber behind the head box leading to two telescopic tubes on either side of the seat which also acted as seat rails.

I was given overall responsibility for these seats and embarked on several visits to Follands at Hamble which had by then become part of the Hawker Siddley Group. These visits were to obtain briefing on the operation and maintenance of the seats, amass all the necessary spares special tools and cartridges required and arrange for the transfer of the seats to RAE. The staff at Hamble were extremely helpful in achieving this and later with our development of the seats for our research programmes.

The Guide Surface Parachute

The guide surface parachute was invented in the USA and taken up by GQ Parachutes of Woking, presumably under licence to develop for use as a stabilising parachute on ejection seats, hence the joint programme with us. The parachute was very good for this role as it was very stable, of very robust construction and had a very low opening shock. The latter meant that it could be deployed very early in the ejection sequence and start its work of stabilising the seat straight away.

My GQ partner in this programme was Arthur Harrison, another larger than life character who in his younger days had been a test parachutist and had even done a live test ejection. All this was now behind him as a prodigious appetite for good food and alcohol had somewhat depleted his fitness. He was a very likable individual and great company.

The first problem we had to overcome was the increased bulk of the guide surface parachute over that of the normal stabilising parachute of the Folland seat meant that it was too big for its compartment in the head box. We overcame this by GQ making larger flaps at the top of the head box to cover the packed parachute. It was still a very tight fit and its deployment was a cause for concern, so we tried packing the parachute in various configurations and doing pull outs in the workshop until we found the best combination.

Having sorted this out we decided that we needed to establish the best attitude to stabilise the seat in. To this end we organised a series of tests in the 24-foot wind tunnel and found that the best attitude was leaning forward at about 30 degrees. In this attitude, there was minimum pitching moment and a moderate positive lift. It was also deemed to be about the best attitude for separation of the pilot from the seat. GQ then set out to design and manufacture a four-point strop assembly to achieve this attitude.

For the trials at Pendine, we obtained a Gnat cockpit and converted it into a sled of similar design to the Lockheed F104 sled but with the retro rockets pointing very slightly downwards after the incident with the latter. We then commenced a programme of trials at Pendine and these went very well, and on completion I wrote a RAE Report on the programme.

Cardington

As well as developing the guide surface parachute with GQ we also used the new GQ aero conical parachute as the main parachute on our trials. As part of the development of this parachute which GQ were lobbying Martin Baker to adopt to replace their Irvin main parachutes, Arthur and I carried out some trials at RAE Cardington near Bedford.

This site was dominated by two huge airship hangars built between 1917 and 1928 to house the R100 and ill-fated R101 airships but at this time housed a number of balloons which were mainly used to test parachutes and train parachutists. These were called barrage balloons by the British and blimps by the Americans. These hangars were later used for the Goodyear airship which was then seen as a cheaper alternative to aircraft for carrying freight but this came to nothing and it was in fact mainly used for aerial filming of sporting and other events.

For our tests, we used the RAE dummy and the normal Folland seat backpack, with the parachute being deployed by a static line rather than the guide surface parachute. The trial consisted of a number of launches from the balloon at different heights and was carried out to Arthur's satisfaction. The balloon crew told us that the main danger in operating the balloons was electric shock from static electricity in electric storm conditions and operators had been hospitalised on a few occasions.

Fluidics

This programme was run by Dave Nosworthy with a team of engineers from Honeywell providing the fluidics expertise but Alf Rivers and I took care of the mechanical engineering aspects of the programme. The then emerging technology of fluidics used small pneumatic devices to control much more powerful devices in a process similar to electronic amplification. In our case, it was hoped that fluidic devices could be used to thrust vector the seat rocket efflux. In the fullness of time, this technology was rendered obsolete by digital technology for most proposed applications.

The Rocket Pack

To do this we needed to have a seat rocket pack which the Folland seat did not have. There was however, a design study by Follands for such a rocket fitted in the seat pan with two nozzles emerging through holes in the back corners of the seat. The Folland drawings were handed over to us and Alf Rivers and his team set to work designing a rocket pack based on this material.

Having designed the rocket casing, we now needed to look at the propellant and for this we approached the Rocket Propulsion Establishment (RPE) at Westcott near Aylesbury. They suggested a bundle of 1" tubes of nitro cellulose to give the 6000lb thrust for 0.25 seconds that we needed. They also gave us a thorough briefing on solid fuel rocket technology. A typical solid fuel rocket had a charge of propellant with a star shaped hole in the middle and the shape of this hole and the quantity of propellant decided the thrust and duration of the burn.

Having designed and manufactured a rocket pack we now needed to test it and RPE offered the services of their outstation at Langhurst, near Horsham in Sussex. This establishment had a building specifically designed for testing small rocket motors and our rocket was bolted to a plate built into the floor with the nozzles pointing upwards. This plate had thrust/time measuring instrumentation built in.

The test firing was successful and achieved the results we required, but the efflux blew the roof off the building which alarmed us more than somewhat. However, the RPE staff reassured us that this was fairly normal and the roof was a specially designed frangible roof which was just there to keep the rain off. When I think of this now, Michael Cain's outburst in 'The Italian Job film' comes to mind. "You're only supposed to blow the bloody door off."

Fluidic System Trials

Having assured ourselves that we had a working rocket pack Dave Nosworthy and the Honeywell team planned a programme to develop the fluidic system. The first thing the Honeywell team suggested was to see if one of their sensors could accurately monitor the motion of the seat on its entry into the airstream and separation from the gun by comparing its output with that of our normal dummy instrumentation. To this end the fluidic sensors were fitted to the side of the seat head box.

This done, a test ejection was set up at Pendine using the Gnat sled with a run at 400kts. However, disaster struck when the sled boost rockets fired and the sled accelerated down the track, the seat rose up out of the sled and crashed down at the side of the track.

When the sled had completed its run and the site had been declared safe, the first thing I had to do was make the seat safe by refitting the safety pins in the sears of the ejection gun and rocket pack. This was a pretty tense episode for me and I felt a bit like a bomb disposal officer.

Examination of the somewhat crumpled seat showed that the gas locks at the top of the ejection gun indicated unlocked which did not really tell us anything as this would be normal after a seat firing. Nevertheless, the seat could not have risen up if the gas locks had been locked and worked properly. Inserting the seat and getting the gas locks to lock was always difficult and Ron and I were sure that the lock indicator was red for lock on both sides. The only conclusion we could come to was either a faulty lock or a false lock indication, although for this to happen on both sides seemed pretty unlikely but perhaps just one lock was insufficient to hold the seat in under the severe acceleration provided by the booster rockets.

I took two actions as a result. Firstly, I stipulated that in future, seat loading into the sled the gas lock should be tested by attempting to pull the seat out with the crane. Secondly, I arranged for the RAF engineering authority for the Gnat to be made aware of the problem. I am not sure of how the fluidics programme continued after that as by then I had moved on.

The Beach Hotel

For all our Pendine trials, we stayed at the Beach Hotel located at the west end of Pendine Sands. The latter's main claim to fame was that they had been the venue for world land speed attempts in the 1920s.

Staying at the Beach Hotel was a long-standing tradition of the Escape Systems Section and I could see why. The landlady was a very friendly, welcoming, local lady who cooked absolutely superb evening meals. Her husband was a retired army sergeant who was working on the ranges when he met the landlady and soon after that they were married and he retired from the army to help run the hotel.

He was quite a character with a 'Pythonesque' sense of humour and, in the evenings, he would regale us with comic fantasy war memoirs including the story of a soldier who did a crash course at the Royal Military College at Cranwell and passed out a corporal. This was followed by a tale of his exploits in a daring wartime raid on a German dolls eye factory.

He also told stories about local characters like Idris the bus and Tudor the undertaker and even one about Mr Templeton, the Head of Engineering Physics Department, who had some time previously attended a trial to witness a test ejection. He stayed at the Beach Hotel with his wife and tiny Chihuahua dog. The landlord's story involved the dog getting lost and a thorough search of the hotel, the village and the beach had failed to find him. When Mr Templeton returned to the hotel for a rest, he got out his curly pipe and on opening his tobacco pouch found the dog curled up inside fast asleep. The landlord told these fantasies in a trance like state in a funny voice a bit like a tribal shaman.

The landlady was also good at telling amusing factual stories about her eccentric relatives and other local characters. My favourite of these concerned her niece, a little girl who was being taught the lord's prayer at Sunday School and 'reinterpreted' one of the lines as 'Lead us not into Tenby station'.

Our daily routine during a trial started with a very hearty full English breakfast. At lunchtime, we returned to the hotel for a couple of pints and a snack in the downstairs bar. At close of play on the range, we would have a wash and brush up followed by a return to the downstairs bar for another couple of pints before going up to the dining room for our main meal with a couple of glasses of wine. After a sojourn watching TV in the lounge, we would go back to the downstairs bar to spend the rest of the night.

After a week or so of this gross overindulgence, we were quite glad to go home to a few days of beans on toast or egg and chips and definitely no alcohol. Surprisingly during all this, I usually found time for a long walk along the beach or over the downs at the west end of the beach. I have always been prone to take long exploratory walks in 'new territory'.

I received some temporary relief from excessive alcohol consumption on the second day of one of our trials. A few weeks before I had a difficult tooth extraction which unknown to me left some fragments of tooth or jawbone still in my jaw. At about dawn on that night, the hole started bleeding and would not stop. After breakfast, George took me to the sick bay where a doctor staunched the bleeding and told me not to have any alcohol for a week or two. I was a bit dismayed at the news at first but was soon relieved to have an excuse to drink just orange juice or gin and tonic without the gin.

The other occasions when I had some relief from heavy drinking were the two instances when I brought Carole and baby James with me. If it was good enough for the head of Engineering Physics, it was good enough for me. So instead of drinking in the bar I was relaxing on the beach with Carole and James.

The main offender in all this overindulgence was Artur Harrison and it was quite normal after our evening meal for him to have the remains of a pint of draft Guinness, a glass of wine and a brandy which he would consume in quick succession to complete his meal.

On one occasion, the discussion during the evening got around to the merits of various wines, mainly Riesling, Liebfraumilch and Neersteiner, German white wines being all the rage at that time. I seem to remember that the main contributors to this discussion were Des Rylands and Jack Gaydon and at one stage they turned to Arthur and asked, "What do you think Arthur?"

Arthur's response to this was, "Don't ask me I'm more of a beer guzzler."

He certainly drank draught Guinness, which had just been introduced to the UK, in prodigious quantities.

On one of our later visits to Pendine, Arthur, having got an enthusiasm for making his own beer, brought some of his homebrew with him for us to sample. It was a very mild tasting beer, a bit like drinking orange juice, but proved to be lethal. On another occasion, he had just returned from being the parachute consultant for the film 'Battle of Britain' and came over all 'theatrical', including wanting to be known as 'Art'.

The Swedes

GQ Parachutes supplied escape system parachutes to Saab in Sweden, who fitted their own ejection seats to their military aircraft rather than using Martin Baker seats. Arthur approached us on their behalf to see if two of their engineers could attend one of our trials at Pendine as they were interested in what we were doing with guide surface parachutes. We readily agreed and they duly attended our next trial.

The two Saab engineers turned out to be similar characters to Arthur, loud and heavy drinkers. They were keen to try British beer but were not too sure about draught Guinness. During one session in the bar one of them spotted the dartboard and wanted to have a go. We explained the rules to him and he took up a set of darts. To everyone's alarm he decided to throw the dart underhand instead of in the conventional manner, causing general movement well away from the dartboard. Surprisingly this technique worked for him and he was soon as accurate as the rest of us, which was probably not saying a lot.

Dylan Thomas and Idris the Bus

The local celebrity was Dylan Thomas who had lived locally at Laugharne. Some of the locals were a bit dismissive of his talents as a storyteller, saying that he did not need any imagination to invent his characters, they were already there in Pendine and Laugharne. They were very upset, however, when the Richard Burton and Elizabeth Taylor film of 'Under Milk Wood' was filmed at Fishguard rather than Laugharne. In fact, the movie was filmed at the nearby estuary village of Abergwuan which was also known as Lower Fishguard.

One such character was 'Idris the Bus'. He ran a bus service between Carmarthen and Tenby and would stop off at selected pubs en-route for a Guinness. On one lunchtime, we were in the bar of the Beach Hotel having a snack and drink when Idris made his usual appearance. After his Guinness, he departed but a few minutes later his head appeared around the door saying, "Give us a push, boys." This was obviously a regular occurrence because the bar emptied and we all pushed the bus down the road until it burst into life.

On a couple of occasions, we finished early on the range and I joined Carole and James on the beach. The top end of the beach was used as a car park and when the tide came in the odd car got stuck in the sand. A local farmer had a 'nice little earner' towing stuck cars off the beach but on one occasion a local rugby team was practicing on the beach and on leaving one of the team member's Mini got stuck in the sand. But instead of calling up the tractor, half a dozen or so team members picked up the Mini and carried it off the beach.

On one of our later stays at the Beach Hotel, a condom dispensing machine was installed in the men's toilet at the downstairs bar. It was not long before the following graffiti appeared on the front of the machine:

'This is the worst chewing gum I have ever tasted'.

Mussels and the Three Titter

On the 'fluidics' trial, one of the Honeywell engineers turned up in a tiny Fiat 500 car armed with a Camping Gaz and suggested that it would be a great idea to go down to the rocks under the headland, pick mussels off the rocks, boil them over the Camping Gaz and eat them. *Why not?* We thought, so four of us squeezed into the Fiat drove over the beach to the rocks and did just that, my first taste of mussels.

It was not unusual for trials to be delayed due to poor visibility or rain and on one of these occasions being at a loose end at the Beach Hotel, the landlady suggested we visit her cousins' farm. Arthur, George Alf and I took up the offer and went there with the landlady. The cousin showed us around his farm and we ended up in the milking sheds while milking was taking place. Arthur pointed out that one of the cows looked a bit under par to which the cousin responded, "Yes she's a three titter."

Moving On

The guide surface parachute programme having come to an end, I was beginning to suffer from wanderlust and spotted a job vacancy for an engineer at the RAE Bedford Airfield. I applied for it and as luck would have it, Geoff Spurr's brother Alan was Superintendent of Engineering Services at RAE Bedford so my transfer to Bedford was quickly agreed and actioned.

Another reason for the ease of my transfer was that the future of the Aircraft Escape Systems Section looked somewhat uncertain. With the imminent opening of the Martin Baker test track, which was a much better facility than Pendine's, the section would no longer be needed to support release to service trials or Martin Baker trials. Also, the only remaining research programme, fluidics, was beginning to look a bit like a dead end.

Not Thurleigh

As usual we had a Ministry funded stay of three days in the area around RAE Bedford to find a new home. The main feature in the landscape we noticed was several large barns bearing the graffiti 'NOT THURLEIGH' written very large on them. At this time, the RAE Bedford Thurleigh airfield was one of the sites being considered for the proposed third London airport, a project which just fizzled out in the end, presumably because it was found that local regional airports like Stanstead, Luton, Southampton and Bournemouth fulfilled the role.

In our house hunting, we first looked at the villages near RAE Bedford like Sharnbrook and Milton Earnest and then the small towns of Rushden and Higham Ferrers to the north. The next day we switched our search to the east and the area around St Neots appealed to us. We eventually found what we were looking for, a three-bedroom detached house at 4, Parkside Little Paxton.

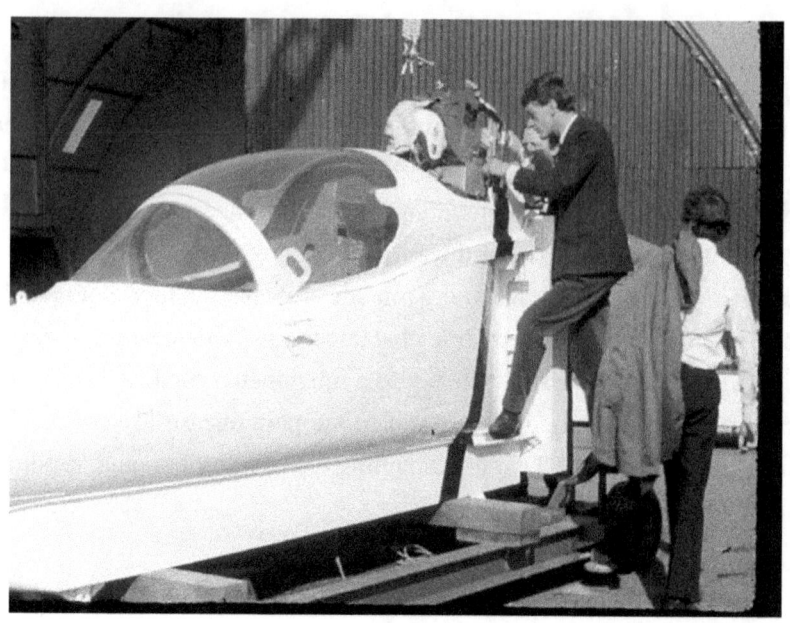

Me, preparing for a guide surface parachute test in the Gnat sled.

The Gnat cockpit sled on the test track.

Chapter 8
Aircraft R&D

1971–1974 Royal Aircraft Establishment Bedford Airfield

1971–73 – RAE Bedford Avionics/Special Projects Engineer

Management of the Avionics Section

Projects included;

Support of stability trials on the Short SC1

Project management of the purchase, refurbishment, conversion and support of the BLEU BAC 1–11

Training/education included;

1. Fracture Mechanics – Sheffield University – May 1972
2. Reliability Maintenance Systems – Cranfield – June 1972
3. Aircraft Performance Assessment – Cranfield – February 1973
4. 4, Management of Human Recourses – MoD – June 1973
5. Operational Research – Bristol University – January 1974

RAE Bedford

On 1 April 1971, we moved into our new house and the following Monday, I reported to RAE Bedford. The establishment consisted of two sites. The main site was the Tunnel Site near the village of Milton Earnest and the site of the wartime Twinwoods Airfield, which was the departure point for the American Bandleader Glen Miller's fateful loss in a flight to France. It housed the main administrative offices, a number of wind tunnels and the main workshops.

The second site was the airfield, near to the village of Thurleigh, this was originally a RAF airfield then a USAF bomber base during the war and in 1954–57 was updated by the addition of a new long runway for the intended National Aeronautical Establishment, this never came to fruition and instead Aero Flight and Naval Air Departments were moved up from Farnborough and the Blind Landing Experimental Unit (BLEU) was moved over from Martlesham Heath in Suffolk.

Engineering Services

This organisation was headed up by Alan Spurr and Included the main workshops on Tunnel site and the staff who ran and maintained the wind tunnels as well as the Aircraft Services Section and the Naval Air Department operation and maintenance staff on the airfield, who were led by Bill Charlson. Ted Payne was the Aircraft Services Manager with myself and Howard Humphries, another junior engineer as deputies. One thing I remember about Howard was the topic of acronyms, which the MoD was very fond of, being used to identify job titles and departments. We were discussing this during a meeting and Howard came up with one for himself. He decided he would be the Electrical and Instruments Experimental Installations Officer or EIEIO.

Ted Payne was only in his post for a short time after my arrival, before he was allocated to a twelve-month management/leadership course at the Royal Naval College at Greenwich. This caused controversy later as Ted thought he had an agreement that he would return to his old post on completion of his college course but when the time came, he was allocated to a post at Farnborough. He fought against this but to no avail.

His replacement as Aircraft Services Manager was Ted Whitley, a retired Royal Navy Lt Commander who I worked with for the rest of my stay at RAE Bedford. I became close friends as well as colleagues to both Bill Charlson and Ted Whitley, exchanging hosting meals and going out to functions. In fact, we continued to exchange Christmas cards with letters attached until the day they died.

Aircraft Services Section

The section operated from three buildings. The main one was the main hangar where the second line maintenance and aircraft modifications took place. Attached to this hangar were a complex of workshops and servicing bays fronted by an office block where I resided. The other two buildings were the flight hangars, one for Aero Flight and one for BLEU.

Under Ted Whitley, Howard was given responsibility for the mechanical engineering aspects and I was given the responsibility for the avionics which included all the electrical and electronic engineering and special projects. I found that managing the avionics sections less daunting than I expected, thanks to a very good electrical supervisor, Ken Balfour.

I thought that I might be resented, as a mechanical engineer, by my avionics staff but this was definitely not the case and my staff surprisingly came to me with their technical problems. Even more surprisingly I was often able to make a real contribution to the technical discussions. My ONC in Principles of Electricity was not wasted after all!

In fact, I came to quite enjoy pouring over wiring diagrams and learning about such ancient and modern equipment as carbon pile rectifiers, rotary and static inverters, inertial platforms, auto stabilisers and autopilots.

Aircraft at RAE Bedford

During the 1955–1994 life of the airfield a lot of interesting aircraft operated from it. During my residence the following were the main ones:

Hanley Page HP115 XP841 and BAC 221 WG722 – These slender deltas were used in the development of Concorde, the former for low-speed handling and the latter, having been converted from a Fairy Delta 2 to the Concorde ogival wing shape, was used for high speed handling. By the time of my arrival, this work was virtually over and they ceased operation in 1970 and 1975 respectively.

Hawker P1127 XP831 and XP984 – These predecessors of the Harrier were being used for assessment of the MADGE landing aid and operation into restricted sites using the nearby Galsey Wood and Twinwoods sites.

Short SC1 XG900 and XG905 – These were being used to explore stability and control during transition and hover. Again, they were coming to the end of their useful life and ceased operation in 1971 and 1973 respectively.

Westland Wessex XR503 – Was mainly used to assess the MADGE landing aid.

Hawker Siddeley HS125 XW930 – The main tasks were investigation of vortex wakes and noise attenuating engine nozzles.

Folland Gnat XP505 – Engaged in the exploration of low level turbulence and cockpit noise.

Vickers VC10 XX914 – Was mainly used for investigation of clear air turbulence.

Hawker Harrier XW175 – This two-seat version was used initially in support of the Sea Harrier programme.

De Haviland Sea Vixen XJ485 – Was used to support Naval Air Department tasks.

Hawker Siddley HS748 XW750 – The mainstay of BLEU Autoland programmes.

BAC1-11 XX105 – The main programmes, after its conversion, were direct lift control and reduced static stability. The former used spoilers to lose altitude or lift without changing the pitch attitude of the aircraft.

Short SC1 SG905

This was the first project I got involved in, mainly to 'play myself in' to my new role. Our main asset for the maintenance and operation of the aircraft's temperamental triplex auto stabiliser was an electrical and instruments charge hand technician called Roy Latreille, and both we and the scientists running the programme relied on him a lot. We did get assistance and advice from a Shorts engineer from time to time, but he was as temperamental as the aircraft and I was told that he ended up going into a monastery. I am not sure whether this was for temporary respite or a more permanent arrangement. We and the scientists gave up the unequal struggle in 1973 and the aircraft was disposed of and ended up in a museum in Belfast.

I think the SC1 was my favourite of all the aircraft I was involved with. It seemed to me to have real character, being small and chunky, cute even. I can imagine it being the hero of a Walt Disney cartoon film with a name like Chuck or Gus.

The Pilots at RAE Bedford

My work at RAE Bedford did not involve close liaison with aircrew but these are the ones I remember;

The chief pilot at RAE Bedford held the title of 'Commander Flying' and the post was filled by either a RAF Wing Commander or a RN Commander. During my tenure at RAE Bedford we had one of each.

Commander Geoff Higgs, the Commander Flying from 1972 to 1976 was a blunt and somewhat distant individual. He was replaced by Wing Commander Hugh Rigg who was a much more approachable person and the brother of the celebrated actress Diana Rigg.

Flight Lt Ron Ledwidge was the main pilot of the SC1 during the period of my involvement with the aircraft. He was quite possessive about the aircraft and understandably somewhat irritated at its low serviceability. The latter was a common problem with new recruits from the Empire Test Pilots School (ETPS). After 12 months of very demanding and intense training at ETPS, the limited flying at RAE Bedford must have been a very frustrating anti-climax.

Terry Downey was the lead pilot of the BLEU Flight and the main pilot of the BAC 1–11. Squadron Leader Peter Gordon-Johnson was the main pilot of the VC 10 during its very short flying career with RAE Bedford. He was later a test pilot at BAE Warton flying Tornado aircraft.

BAC 1–11 XX105

BLEU needed a modern airliner for the forthcoming research programmes on their books, and had decided on a BAC 1–11, several of which were for sale at the time. The aircraft finally chosen was G-ASJD which was being offered by British Caledonian Airways and I was tasked with the project management of its purchase, refurbishment and conversion. This aircraft had a history, in that it had made a heavy landing on Salisbury plain while undergoing certification trials by the manufacturer. However, it was deemed to offer the best value of those on offer.

The negotiations with British Caledonian were at Gatwick with an individual known to one and all as 'Curly' Walter who operated under the title of Special Projects Manager. He proved to be a suave sophisticated individual who somehow put me in mind of a James Bond villain. These negotiations cantered on the spares package, aircraft publications and ground crew training. We decided that we would only purchase a spare Rolls Royce Spey engine and seek a separate spares support package based on a unit exchange scheme. A full set of aircraft publications was agreed and we were pleased to find that the civil publications were structured essentially the same as the military ones we were used to but with different names. British Caledonian agreed to provide us with the full ground crew training at their Gatwick facility.

On the due day of aircraft handover, me and my team travelled down to Gatwick accompanied by two individuals from the MoD Procurement Executive Contracts Branch. I signed on behalf of MoD (PE) and Curly Walter on behalf of British Caledonian.

At the instant, I appended my signature my deputy piped up, "You know he only has signing powers for up to £25," which caused some consternation but the contracts branch representative reassured us that it was OK.

That done we boarded the aircraft for its delivery flight to BAE Hurn where it was to undergo a refurbishment and repaint in a somewhat garish yellow, red and white colour scheme devised by BLEU. This paint scheme caused some controversy as it should have been the standard MoD (PE) colour scheme, popularly known as the raspberry ripple. BLEU though, were determined to have their own individual scheme, so we got the 'banana split' paint scheme instead. An air hostess in a very smart tartan uniform and refreshments were provided for what was a very short trip.

While the refurbishment was taking place Rex Buckley, the BLEU Hangar supervisor and his team of chosen BAC 1–11 technicians were despatched to Gatwick for their training courses where they were accommodated in a B&B in the quaintly named village of Peas Pottage. Meanwhile I drew up a specification for spares support on a unit exchange basis which was put out to tender. We received responses from BAE, British Caledonian, Court Line and Aer Lingus. Of these only Aer Lingus really met our requirements and were also by far the cheapest.

Aer Lingus and Dublin

Before making a final decision, we decided to take a look at the Aer Lingus facilities in Dublin and to this end Ted Whitley and I drove down to Heathrow and boarded an Aer Lingus Boeing 737 for Dublin. This flight was a bit strange as the only passengers were Ted and I and about 30 nuns. We were met at the airport by Patrick Michael Murphy, who was the senior avionics engineer and a colleague who dealt with the contracts side. Murphy, like most of the engineers at Aer Lingus received his training in the RAF.

We were settled into a hotel at the airport and had a meal before being collected by Murphy and his colleague for a tour of the sights of Dublin and a couple of pints of Guinness at a Mooney's Bar alongside the River Liffey. We were shown around the engineering facilities the next day and we were quite impressed by them and the expertise of the engineering staff.

That was the easy bit as persuading the MOD (PE) Contracts Branch to deal with a company from the Republic of Ireland at the height of the 'troubles' in Northern Ireland was not proving easy but they relented in the end. We also had problems with the MoD Police gate guards on the first visit by Aer Lingus to RAE Bedford and we had to ensure that they were escorted at all times. Nonetheless we finally had our spares support contract and Aer Lingus provided an excellent service for the duration of my stay at RAE Bedford.

Cranfield

On completion of the refurbishment at BAE Hurn, the BAC 1–11 was transferred to Cranfield Institute of Technology for its conversion to the BLEU role. The College of Aeronautics at Cranfield, as well as being a university of the air had a very competent engineering facility with considerable experience in the modification of aircraft and RAE and A&AEE used it on quite a number of occasions.

The conversion included:

1. Replacing the Elliot autopilot with a duplex Smiths SEP5 autopilot (SEP was an acronym for 'Smiths Electric Pilot', which sounds rather quaint now).
2. Ferranti inertial navigation.
3. BLEU versatile autopilot.
4. An area navigation system.
5. Digital and analogue computers.
6. Digital recorder.
7. Electronic display console.

During this period, I attended a couple of specialist courses at Cranfield, which allowed me to monitor progress of the BAC 1–11 conversion.

The first of these was 'Reliability Maintenance Systems' which looked at a new maintenance philosophy which was being introduced in the aircraft industry which planned to replace scheduled hierarchical maintenance lay ups with maintenance based on condition monitoring with such information gathering techniques as pilots' reports, non-destructive testing (NDT) and fluid sampling. The philosophy behind these systems could be summarised by the headline; 'If it isn't broke don't fix it'.

The second was Aircraft Performance Assessment, which covered such topics as flight-testing methods, instrumentation and ground test facilities such as wind tunnels. During these courses we had our meals in the senior staff mess and the main thing I remember about this facility was the vast array of whiskies on display behind the bar.

One of the hangars was used as a museum and the exhibits included a TSR2 and a German WW2 V1 pulse jet flying bomb and V2 missile. I believe these exhibits have long since been dispersed elsewhere.

Support for the BAC1–11

While the refurbishment and conversion were progressing, I was looking at other aspects of support for the aircraft. Support for the Spey engines was helped by our resident Rolls Royce representative, Harry Shaw who put us in contact with an engineer at the East Kilbride factory where Spey engines were manufactured. We decided to take advantage of their 'power by the hour' arrangement where a fixed annual payment secured the spares support, updates and modifications we needed.

We purchased the special jigs, fixtures and tools we needed from various sources and were pleased that the major expenditure of an electrical ground power rig was not needed as the aircraft had its own auxiliary power unit (APU).

One expensive item we did need though was equipment to gain access to the high tail plane which was mounted above the fin. We looked at the solutions used by other operators, including a tail dock built into the hangar doors by Court Line at their Luton Airport facility for their Lockheed Tristar aircraft. In the end, we opted for a Simon Hoist for daily inspections and an aluminium tube tail dock for second line servicing supplied by a specialist manufacturer based at Andover.

We soon discovered that a high tail plane aircraft like the BAC1–11 was a very attractive birds' nests site. Birds would have a half-finished nest installed overnight so our first daily inspection during springtime included a lookout for birds' nests. While on this subject, one of the more unusual residents of our main hangar was a little owl who sat on one of the roof trusses peering down at the activities below, attracting a lot of attention in the process.

Our Engine Loan Business Opportunity

After about 18 months of operating the BAC1–11, I had a phone call from Curly Walter requesting the loan of our spare Spey engine to rescue one of their BAC1–11s stranded at Gibraltar. Ted and I agreed we would do this but had no idea what to charge, so we consulted Rolls Royce and Aer Lingus who advised us on the going rate in the civil aviation industry. We then contacted our contracts branch to set it up and at first, they came up with a rate using some Treasury formula which was far lower than that above. We persevered however and in the end our rate was agreed and British Caledonian were only too happy to pay it.

The BAC 1–11 Operators Seminar

Our operation of the BAC 1–11 was helped considerably by a BAC 1–11 operator's forum held annually at one of the Heathrow hotels. I attended two of these with Ted Whitley on the first occasion and Rex Buckley on the second. This proved to be useful in learning and discussing current issues with the aircraft and making contacts with other operators.

BAC used to host this event and they would provide their advice on forthcoming modifications and operating procedures, and invite operators to air their problems and experiences. We were pleasantly surprised at how interested the other operators were in what we were doing with XX105 and our experiences with the aircraft.

Courses

As well as the two Cranfield courses mentioned above, I attended three other contrasting courses in this period. The first was called 'Fatigue Failures and

Fracture' and was held at Sheffield University. It was on the new science of fracture mechanics and the main lecturer was, I seem to remember, Professor Tuplin, though I may have him mixed up with someone else. He was a leading authority on the subject who had written several papers and books on the topic. Consequently, the course attracted a lot of interest and there were students from Europe Japan and the USA attending our course.

Unfortunately, he proved to be a very poor lecturer on this occasion whose presentation skills were zero and whose mumbled cursory response to anxious questions brought dismay to the students. The day was saved by very good supporting lecturers, some good films on the topic and excellent comprehensive course notes written by the said professor.

Sheffield proved to be a surprisingly pleasant city with lots of nearby beauty spots to the northwest which I visited with a colleague from the Tunnel site.

Management of Human Recourses

The second course was a MoD course called Management of Human Resources. It was held at the Beach Hotel on the seafront at Littlehampton in Sussex which was a pleasant if somewhat quaint hotel in a building which was both quirky and gothic, reflecting its chequered history. It was originally built in1775, a first floor added in 1818 and further extended in 1887. I discovered on a much later visit to Littlehampton that it had been demolished and replaced by residential apartments. Another example of local authority vandalism perhaps, like the Worcester City Council demolition of Elgar's last home to build a new housing estate on the land.

The course was a bit strange and cantered on topics like staff motivation and teamwork skills. We were set up in rival teams to compete with each other on rather silly projects like building a paper tower. I was the only engineer on this course, the rest being administrative staff. Some of the ladies on the course found some of the more confrontational aspects a bit distressing and one lady burst into tears at one stage. The men largely took the course a lot less seriously, some to the extent of seeing it as being a bit of a laugh which further infuriated some of the ladies. I suppose it could have been worse, we could have been navigating assault courses in a wood somewhere, with an RSM type instructor shouting at us.

Littlehampton proved to be a very pleasant seaside town with a long promenade and a harbour at the mouth of the River Arun. At the former, I even went jogging on a couple of occasions with two fellow students who were keen joggers. At the latter, we came across an old sailing boat undergoing restoration with a very interesting and knowledgeable owner on hand to tell us about the boat's history and show us around.

Bristol

The final training course I took part in during this period was Operational Research at Bristol University. I don't know why I was allocated for this course as it was nothing to do with any work I was involved in. The subject was a series of optimisation techniques using mathematical modelling.

In spite of that, it proved to be a very enjoyable week located in a mansion called Burwells, alongside the Avon Gorge and Brunel's magnificent suspension bridge. I believe that the mansion belonged to the Wills cigarette family who donated it to the university. According to Google Earth it is now luxury apartments.

In the front of the house was a large lawn, equipped with a croquet set and we played our version of the game during tea and lunch breaks, none of us knowing the rules of the real game. My evenings were spent going for walks in the nearby Leigh Woods and Clifton Down with its Camera Obscura Observatory and spectacular viewpoints.

The Mini Jem

One day, while scanning car adverts in Exchange and Mart, I came across one for a Mini Jem kit. This was a fibreglass coupé which used Mini components. I rang the number and found that the man, who lived near Silverstone motor racing circuit, had purchased all the parts necessary to build the car but found that he had neither the skill nor time to complete the task. We went to view the kit and I agreed to purchase if he could deliver it.

A few days later the kit arrived on a trailer and was unloaded into our garage. For the next few months, I spent most evenings and weekends in the garage assembling the car much to Carole's annoyance. The assembly was fairly straightforward except for the gearbox, which was a Colotti 5 speed racing gearbox, which came as a box of bits and it took some time for me to work out how it all fitted together. Fitting the front laminated windscreen also proved to be something of a nightmare and I cracked the screen on my first attempt and had to purchase a replacement from the manufacturer based at Cricklade in Wiltshire. They provided some useful advice on the fitting of the windscreen and a very useful special tool.

When completed, the car proved to be sprightly with good road holding but extremely noisy inside and the non-synchromesh gearbox proved very difficult to select the right gear in town traffic. Carole did not like the car at all and refused to go in it or let James go in it either. After about nine months, I sold the car for a profit.

The sale of the car turned out to be a bit bizarre. I advertised it in Exchange & Mart and received an almost immediate enquiry from a chap who agreed to come and view the following evening. Unfortunately, going to work the next day I had an accident in a sudden bank of thick fog which resulted in a crumpled nose on the car with bits of fibreglass sticking out of the cracks but no damage to the mechanical components. When I got to work, I telephoned the prospective buyer to tell him about the accidental damage to the car. Strangely this did not seem to put him off and he still wanted to come and see the car. That evening the chap duly turned up, looked the car over and paid the asking price in cash.

He lived in North London and I was a bit concerned about him driving down the A1(M) with the car in that state so I rang him a couple of hours later to see if he had got back all right. He answered the phone in a state of high excitement saying, "I nearly got a ton out of it on the dual carriageway."

Hi Fi

I had amassed quite a collection of jazz records in the late 1950s and early '60s but by now had nothing to play them on. To remedy this, I scanned Hi Fi magazines for guidance on suitable budget systems. I also found a Hi Fi dealer in Bedford who provided lots of useful advice and demonstrated selected systems to me.

This resulted in me purchasing a turntable, cassette deck and amplifier from them. The turntable was a Pioneer PLD 12, which was the most highly regarded budget turntable of the time. I cannot remember the make and type of the other two items but can remember what they looked like.

For speakers, to save money and still have good quality, I purchased a Wharfdale kit for a pair of speakers which comprised of the drive units, the components for the crossovers and plans/instructions for the cabinets with the option of purchasing cabinets from a specialist manufacturer.

I decided to go my own way and designed and built a pair of cabinets comprising of an outer frame of 1inch square softwood and top, bottom, front, back and side panels in 5/8inch mahogany board which I obtained from Dad's hoard of timber. I was very pleased with the appearance of the completed speakers and when I hooked the system up and put an LP on the turntable, I was amazed at the sound quality. Later, after a number of system upgrades, I gave the system to Carole's mum.

I continued to read selected Hi Fi magazines, which started to rave about a combination of the Linn Sondek LP12 turntable from Glasgow and amplifiers from a small firm in Salisbury called Naim Audio. Their philosophy was to strip the system of all unnecessary frills and use the best quality materials and components. This minimalist approach very much appealed to me and I was 'hooked'.

I purchased the Linn turntable and Naim amplifiers from Grahams Hi Fi, a dealer located near Kings Cross Station over several business visits to MoD offices in London. Carrying the turntable in its huge box through the streets of London and on the train from Kings Cross to St Neots Station was something of a challenge. I later became more acquainted with Naim Audio after we moved to the Salisbury area.

Promotion

At the end of the BAC 1–11 conversion programme, I was invited to attend a promotion review board at one of the MoD (PE) offices in central London and was successful. Because of this Alan Spurr decided to create a new post at main grade engineer level for me to fill. It was titled Aircraft Installations Manager and I would work alongside Ted Whitley with me managing aircraft modifications and Ted aircraft maintenance.

Short SC1 XS 905 hovering over Galsey Wood. This picture is of the aircraft in its earlier pre-crash and rebuild state. -

BAC 1-11 XX 105 on a guided approach to Thurleigh. I don't know what the equipment in the foreground is, presumably a RAE experimental approach aid.

The Mini Jem at Little Paxton

Part 5
Senior Engineer

Chapter 9
Airfield Rationalisation

1973–1977 Royal Aircraft Establishment – Bedford

Aircraft Installation Manager

Management of the facilities for the design, manufacture and installation of experimental equipment in aircraft.

Projects included;

1. Project management of the purchase, refurbishment and conversion to a flying laboratory role of a Vickers VC10 aircraft.
2. Folland Gnat instrumentation update.
3. Planning and co-ordination of the transfer of the RSRE Pershore aircraft and facilities to RAE Bedford.

Aircraft Installation Manager

There were two less than ideal aspects to my new role. The first was that I did not have a design office of my own but relied on the Aero Flight and BLEU design offices. This would be rectified after my departure when we inherited the RSRE Pershore design office. The second was that Ted and I shared some of the hangar staff. This did not prove to be a problem as Ted and I got on very well.

Vickers VC 10 XX 914

In 1973, a case was made for a large transport aircraft for research programmes by Aero Flight and Structures and Engineering Physics Departments at Farnborough. The main areas to be explored were structural response to atmospheric turbulence, including clear air turbulence, high speed buffet and vortex wakes. To this end a VC 10 was purchased, again from British Caledonian. With the experience of the BAC 1–11 behind us, the purchase was much more straightforward on this occasion. This time the delivery flight was to Filton where a refurbishment programme and comprehensive strain gauge fit was undertaken by BAE.

During my visit to Filton, mid programme, the establishment appeared to be having a very quiet period post Concord. The VC 10 was the only aircraft in a very large hangar and there seemed to be an air of despondency about the place.

The product support for the VC10 was located at Weybridge and a visit there revealed more gloom and despondency. By this time, Weybridge was being run down with staff being transferred to Filton and a shuttle fleet of coaches doing a daily run to Filton. The VC10 product support appeared to fall well short of the excellent support provided for the BAC 1–11.

This time support of the aircraft was very straightforward as it could be taken care of by the RAF as their VC 10 fleet workload was reducing. A visit to RAF Abingdon, where engineering support for the RAF VC 10 fleet was located, confirmed this with the engineering staff there being very willing to provide the help and advice we needed. I also visited RAF Brize Norton, where the RAF VC 10s were operated from, with Ted Whitley to look at flight servicing and ground crew training aspects.

The aircrafts service life at RAE proved to be short however as the aircraft was grounded during a round of defence cuts in 1975 after a failed clear air Turbulence trial at Vancouver Island in Canada. The reason for the trial's failure was that, very exceptionally, no clear air turbulence occurred there that spring. The aircraft languished between the main and Aero Flight hangars for a number of years, being cannibalised for spares by the RAF, before being broken up for scrap.

Folland Gnat XP 505

In June 1976, the aircraft was laid up for an instrumentation update for research programmes investigating low level turbulence and cockpit acoustic noise. The most radical modifications we planned were the fitting of long probes on the nose and both wing tips, so we felt we needed the advice of the aircraft design authority. Also, we intended to fit even more recording equipment in the rear cockpit, rendering the aircraft a single seater. By this time, Follands had been part of the Hawker Siddley group for some time and the latter's amalgamation with BAE was imminent.

On investigation, we found that the Gnat design authority and records were now held at the Kingston on Thames factory so we arranged a visit to discuss our proposals. Our delegation included myself, John Cannell, the Aero Flight project manager and a designer from the Aero Flight design office.

On arrival at Kingston, we found that although the design records were indeed held at Kingston, there were none of the original Folland design team left and the Kingston design team were not terribly interested in the Gnat, being heavily committed to their new advanced trainer, the Hawk.

On the return journey to Bedford, to alleviate our frustration and disappointment, John suggested we visit Kew Gardens and I readily agreed, having spent a couple of very pleasant visits with Carole and James when we lived at Bordon during my Farnborough period.

In the event, the comprehensive instrumentation update was not completed until October 1978, after my departure.

The Farnborough-Bedford Air Ferry

During this period, I made use of this facility several times. The facility consisted of four De Haviland Dove aircraft based at Farnborough and ran scheduled trips between Farnborough and Bedford. Trips to RAE Aberporth, Llanbedr and West Freugh were also carried out on a request basis.

My trips on the ferry were to attend meetings at Farnborough and would involve return trips in dusk or darkness between November and March. My main memory of these is the burning of wheat stubble by farmers after harvesting which was prevalent at the time. This provided a spectacular view from above. This practice was later banned.

One of these trips concluded with a very spectacular and frightening incident. It was on an afternoon return trip in summertime. On the approach to the airfield, we were about to touch down when the aircraft leaped back into the air and wobbled furiously. After a safe landing, the pilot admitted that he had lost control and we were lucky to touch down straight and level. The cause of this disturbance was the vortex wake of a VC 10 which was doing 'touch and go' circuits. At this time, Thurleigh Airfield was used by BOAC and BEA to train pilots.

The Airfield Rationalisation

In the latest defence review, it was stipulated that one of the four main MoD (PE) airfields had to be closed. It was deemed that the airfield at Boscombe Down was essential and was excluded from the review, so Bedford, Farnborough and Pershore were to be reviewed to see which one was to be closed.

The review consisted of the three establishments making a presentation to a committee who would decide our fates based on this and other evidence gathered. I was nominated to do the presentation for RAE Bedford and started to compile a case for Bedford based on our engineering facilities, our record on aircraft support, our procedures for the design, manufacture and installation of aircraft modifications, the airfield facilities and the strategic advantages of the airfield location.

On the due day, we all turned up at Farnborough for the presentations and I found mine to be a bit nerve racking but it seemed to be well received. The Pershore presentation was given by Bill Sleigh, who had replaced David Henderson as their Chief Engineer. Pershore had the most impressive record for aircraft modifications and the best developed system of documentation for recording and clearing them. Unfortunately, Bill overstated this case and laboured it. Also, the Pershore airfield facilities were hardly touched on and were in any case not as good as Bedford's. The Farnborough engineering facilities and procedures were the same as ours and their presentation was a bit cursory, as if to compensate for the lengthy Pershore one. Farnborough probably had the weakest case as far as the airfield was concerned with restrictions due to the built-up areas all around and the nearness of Heathrow.

Having said that, Farnborough is the only one of the airfields still operating today, albeit as a commercial airfield operating business and executive jets. The merits of the presentations aside, we thought that Bedford had the strongest case and Farnborough the weakest and were surprised when the decision was made to close Pershore and relocate the aircraft and facilities to Bedford. Another possible reason for the decision was that Pershore did not appear to get much support from their parent establishment at Malvern, who seemed almost to want to be rid of them.

The transfer of RSRE Pershore

Almost as soon as the announcement was made, we were invited to Pershore to discuss the transfer of aircraft, facilities and staff to Bedford. For this, the first of three visits, I was accompanied by Bill Charlson, who by then had reached retirement age but had been retained in a lower grade by the simple expedient of swapping jobs with Ted Whitley.

On the journey to Pershore, as we came over Broadway Hill, the view that met us was of the Vale of Evesham and the Severn Valley being shrouded in mist with only the Malvern Hills and Bredon Hill being visible as islands in a sea of mist.

I had feelings of trepidation about meeting old colleagues, including my old boss, Harry Cameron, the head of the design office, in these circumstances, but we were surprisingly very well received.

The first thing we established was that very few staff would be coming our way as most had either found alternative employment at Malvern or opted for early retirement. This included Bill Sleigh, the Chief Engineer and the Aircraft Servicing Manager who had been offered jobs at Malvern and the nearby MoD logistics facility at Ashchurch respectively and Harry Cameron, who had opted for early retirement. We did though, talk to a few staff who were in fact quite keen to come to Bedford.

We soon established a list of aircraft which would be transferred and the facilities like special tools, spares and ground support equipment which would need to be transferred. From this information, we established that the two ex-Naval Air Department T2 hangars and associated outbuildings would be sufficient to house the RSRE air fleet. It was also agreed that more general equipment like vehicles and machine tools would not need to be transferred.

One exchange at our first meeting at Pershore lingers in my mind. The discussion had got around to the removal of heavy equipment for transportation to Bedford and the need to arrange a contractor to do this. At this stage, the chairman of the meeting said, "Why can't the heavy gang do this?"

To which Harry Cameron retorted, "Have you seen the heavy gang, they are all cripples."

Our visit concluded with a very pleasant meal in the officers' mess with Bill and Harry.

During this visit we stayed the night at the Manor Hotel at the south eastern end of Pershore, near to the bridges over the River Avon and the meadows where I proposed to Carole. The hotel proved to be a bit disappointing, seedy even, with an air of faded elegance compared with the visits we made during our courtship. That evening I made a visit to my parents at nearby Spetchley and took Bill along with me as he did not relish spending the evening alone at the hotel.

Our next meeting was, in fact, at Malvern with Airborne Radar Department to see what facilities their team at Bedford would need. Here, the first thing we learned was that most of the scientific and engineering staff being transferred to Bedford would in fact be contractors from the radar companies like Ferranti. Their main building requirement would be for a tower building located on a high point within the airfield site and it was agreed that a RSRE team would come to Bedford to find a suitable site.

The RSRE team duly arrived and chose a location at the highest point at the north end of the site, in the middle of Galsey Wood. This immediately posed a problem as Galsey Wood had been a bomb dump tor the USAF during their tenure in WW2 and had never been formally cleared. A full-scale survey by a specialist Army team was arranged and all in fact that they found was a few rounds of light ammunition.

My final involvement with RSRE was a trip to the Royal Navy Air Station (RNAS) St Mawgan in Cornwall to view some Handley Page Jetstream aircraft which were becoming surplus to RN requirement after having been used as avionics systems trainers. RSRE were considering putting in a bid for one for use as a laboratory aircraft and sought our advice on the aircraft's viability. On the due day, Ted Whitley and I drove to Pershore to be flown down to St Mawgan by RSRE.

Our transport turned out to be a Puma helicopter piloted by an enthusiastic young RAF pilot based at Pershore. The first thing he did after take-off was to fly low over his house at nearby Lower Moor and wave to his wife who was in the garden. We were made very welcome at St Mawgan and soon established that all the support we needed to operate the aircraft was readily available. On the return trip, we flew low and fast over Dartmoor causing cows to scatter. In the end, these Jetstream's were not procured by RSRE.

Cornwall

In our last two years at Little Paxton, we had our first proper holidays since our honeymoon. Our friends Sylvie and Brian recommended a B&B at Rescassa Farm near Mevagissey where they had two very enjoyable holidays. It was indeed a very good B&B with a very hospitable farmer's wife and superb meals of local produce including on one occasion John Dory fish. The farm had a prize herd of Guernsey cows and the fresh milk from them was delicious. One morning we even got up early to see them being milked before having breakfast. The nearest beach, Porthulney Cove was a very attractive and secluded beach and across the road was the local mansion, Caerhays Castle, which looked like the location for a Daphne du Maurier novel. I believe it was used for the TV adaption of her novel Rebecca.

On both occasions, we broke up the journey down by spending the first night at a B&B near Oakhampton. There was a pub near the B&B which featured jazz played continuously on a Hi Fi system. There were also pictures of jazz musicians all over the walls. It was a very atmospheric pub with low ceilings, flagstone floors and a labyrinth of small alcoves on different levels. The landlord gave every appearance of being high on drugs but he seemed to deal quickly and efficiently with bar transactions. While at Oakhampton, we also visited the castle and Yes Tor on Dartmoor.

On arrival at Rescassa Farm, we decided to spend our first afternoon on the beach at Porthulney Cove. After we had been on the beach for around two hours with James building sand castles, the beach had become quite crowded. I casually suggested we build a barricade to stop the tide coming in and started work on it. Carole and James soon joined in, building a deep ditch and embankment parallel with the incoming tide. We had been at this some time when I noticed that the family next to us had started to build their own barrier which soon joined on to ours. After about an hour, the barrier extended half way along the beach, executed by all the family groups along the tide line. By then, it was approaching the evening meal time agreed with the farmer's wife so we left and did not see the inevitable engulfing of the barrier by the tide.

During our first holiday in August 1975 we restricted our itinerary to local beaches and attractions, Including Gorran Haven, Mevagissy, Boswinger, Portholland, Portloe, Veryan and Portscatho. All these destinations involved very narrow, hilly, twisty roads for which our then car, a Citroen Dyane was ideally suited. One day I decided we would try a cliff walk so I parked in the car park overlooking Porthulney Cove and set out for Portholland. On the Ordnance Survey map, it did not look far but it was very twisty and hilly, so by the time we reached our destination Carole and James were complaining bitterly.

We had one rainy day so after a visit to a tin mine attraction we decided to try the north coast and went to St Agnes, which was bathed in sunshine. During both holidays, for lunch we usually went to a pub or café for a light snack. I remember one occasion looking at a menu of sandwiches and James chose lobster, which was the most expensive choice, and he could not be persuaded otherwise. I expected him to take one bite and decide he did not like them, so I was looking forward to swapping and enjoying the lobster, but he ate the lot and said he enjoyed them.

On our second holiday, the following year we ranged further afield, visiting Charlestown Historic Harbour, Chyrauster iron age village, St Michaels Mount, Truro and King Harry Ferry where there were a lot of redundant oil tankers parked up on the estuary. We also visited St Maws and took a boat trip up Carrick Roads estuary.

This time there was a couple from Bradford with a son about the same age as James staying at Rescassa Farm who we became friendly with. On a couple of evenings, we all walked over the fields to the pub at Gorran Haven and on the return journey I spotted a bull in one of the fields of cows. I said nothing to the others and the bull ignored us. I was quite relieved when we crossed into the next field and I told the others.

On our return journey, we stopped at Launceston and looked around the castle. We also broke up the return journey by spending the night at a B&B farm near Polsham, north of Glastonbury. The next day we explored Glastonbury, climbed up to the Tor and explored Wells and its cathedral before going to spend two days with our parents in Worcestershire and finally returning to Little Paxton.

Transfer to Boscombe Down

One day I was browsing through the MoD vacancy notices and spotted one for a squadron engineer at A&AEE Boscombe Down. I am not sure why now but I found this vacancy very appealing and mentioned it to Ted Whitley who also thought it would be a good job for me, having during his royal navy career been The Engineer in Charge of C Squadron, which was then the naval aircraft squadron.

I therefore applied, was interviewed at Boscombe down by the Superintendent of Technical Services and Aircraft Servicing Manager who decided they were quite keen to have me. As this was a level transfer rather than a promotion, it should have been very straightforward but RAE Bedford would only agree to release me when a suitable replacement was found. The prospects for this did not look good, so I was becoming increasingly frustrated by the impasse.

However, this was resolved when a junior engineer on the tunnel site applied for my job and this was agreed with a suitable handover from me. Having found a suitable house at 13 St Just Close Newton Tony, just 4 miles east of Boscombe Down, we moved in on 1 April 1977 and I reported for duty at Boscombe Down on the following Monday.

CAROLE of Portscalho 197

Badminton on the beach with Carole.

Chapter 10
Helicopters

1977–1981
Aircraft & Armament Evaluation
Establishment (A&AEE) Boscombe Down

Engineer i/c D Squadron
Management of the maintenance and preparation for trials and aircrew training of all helicopters at Boscombe Down
Projects included:

1.	Preparation and support of Lynx oversees and ship trials.
2.	Organisation training and equipment procurement for Chinook heavy lift helicopter.
3.	Management of a comprehensive flight test instrumentation and special equipment fit in the Chinook.

Training and education included:

1.	Flying training for PPL – Thruxton Airfield – 1977/8
2.	Lynx Helicopter Systems – Westland's – 7/78
3.	Electronics for mech. Engineers – Richmond College – 6/79
4.	Design Engineering Conference – MoD – 10/81

Boscombe Down

My next 25 years at Boscombe Down accounts for 53% of my career but I will be hard put to make it 30% of this saga. In fact, my Boscombe Down career proved to be not as interesting, rewarding or enjoyable as my previous career with mostly staff management rather than hands on engineering. It was also marred by a family tragedy and various family traumas.

At the time of most of my career at Boscombe Down, the establishment was headed by the Commandant, who was a RAF Air Commodore with a civilian Chief Superintendent under/alongside him. This depended on the personalities of the individuals concerned, sometimes one was dominant and sometimes the other. This impression of an establishment with two heads was always something of a bone of contention. The Chief Superintendent headed up the assessment divisions which included, Performance, Engineering, Armament, Navigation and Radio and Trials Management. The Commandant headed up Flying, Technical Services and Administration Divisions.

Technical Services Division

When I reported for duty at Boscombe Down, I was first of all interviewed by the Superintendent of Technical Services Division, John Bennett and the Aircraft Servicing Manager Dave Wilson. The first thing John Bennett said was that as a squadron engineer, I should do two things. The first was to embark on a flying training course leading to a private pilot's licence (PPL) which Boscombe Down would fund and the second was to become a member of the Officers Mess. I was more than happy to comply with both of these so when the interview terminated Dave Wilson took me to E Squadron to introduce me to my new colleagues.

Technical Services Division at that time consisted of the four squadrons who carried out the maintenance and preparation for trials of the Boscombe Down air-fleet under the Aircraft Servicing Manager who was also responsible for the armament and safety equipment servicing bays. The other half of the organisation, under the Specialist Services Manager included the Design Offices, the Mechanical Workshops, the electrical and electronic servicing bays and Avionic Rigs.

Engineer i/c E Squadron

Although not mentioned in the CV above, for my first six months at Boscombe Down, I was Engineer i/c E Squadron. My new charge was responsible for the Empire Test Pilots School (ETPS) fixed wing aircraft, the Comet laboratory aircraft, a Hunter which carried out spray runs for chemical defence exercises over Porton Down ranges and the three Harvard photo chase aircraft. The squadron operated from two hangars designated E1 and E2. My office was on the ground floor of the office block at the front of E1 whose first floor was occupied by ETPS.

The other three squadrons were A Squadron who were responsible for the fast jets, B Squadron, who were responsible for the other fixed wing aircraft and D Squadron who were responsible for the helicopters.

My deputy, the Principal Supervisor was Doug Young, who bore a remarkable resemblance to my father-in-law, George, both in appearance and even more in their Wiltshire accent, to such an extent that I would have been hard placed to tell them apart. He was very welcoming and imparted lots of useful information on local places of interest and pubs who did good meals.

The Hunter Intake Repairs

My six months in this post were relatively uneventful except for major repairs to the engine intakes of both ETPS T MK 7 Hunters. Serious cracks were found in the intakes of both hunters during a routine inspection. This was a known problem with Hunters and there were well-established repair schemes available.

There was some urgency in getting the repairs done as the Hunters were essential to the ETPS programme as the only aircraft capable of inverted spinning. To this end, Mechanical Workshops having offered their help, we decided to set up two repair teams, one from the hangar and one from Mechanical Workshops. We allocated the most seriously damaged aircraft to Mechanical Workshops as they had the skilled sheet metalworkers.

The rivalry between the two teams helped the progress of the task but one had to be careful not to turn it into a race and thus endanger quality. Nevertheless, the repairs were carried out to a reasonable timescale and lasted the rest of the aircraft's lives.

Transfer to D Squadron

In September 1977, the Engineer i/c A Squadron, Owen Hudson was transferred to MoD (PE) headquarters and the Engineer i/c D Squadron, Les Spicer was transferred across to fill the post, leaving his post vacant for me to fill.

With my limited knowledge of helicopters, I decided that I needed some sort of helicopter training. Fortunately training courses on the new Westland Lynx helicopter were becoming available at Westlands, so I applied. The course was not at the factory at Yeovil but at a newly established training centre at the south east side of Sherbourne, a very pleasant location alongside a stream and with a view of Old Sherborne Castle.

The Lynx boasted three new features. The first was a semi rigid rotor which had far fewer moving parts than a conventional articulated rotor and gave the aircraft a higher speed and greater agility; it was the only helicopter which could do a roll manoeuvre. This came at the expense of greater noise and vibration and I well remember a later trip to Cranfield for a meeting in a Lynx and being rendered deaf for the first 30 minutes or so of the meeting.

The second was a main rotor gearbox utilising conformal gears which have area rolling contact instead of line contact, enabling greater loads to be transmitted, which in turn meant fewer gears and a shallower, lighter gearbox.

The third, on the Navy version, was a harpoon under the cockpit which could pierce and lock onto a grid on the flight deck, firmly attaching the helicopter to the ship. In other respects, the Lynx employed conventional helicopter technology.

Flying Training at Thruxton

My flying training started in March 1978 at Western Air Training which operated from Thruxton Airfield near Andover which also doubled as a car and motorcycle racing circuit. My usual instructor, Barry Dyke proved to be an excellent instructor, even if he looked more like a hippy than a conventional flying instructor, with his long hair and scruffy jeans. The aircraft we used were Cessna 150 and 172s and I was surprised to find that there was quite a difference in the handling of these near identical appearing aircraft. The 172 seemed to handle like a lorry after the far more agile 150.

I enjoyed my flying except for spin recovery, which I found to be very alarming. This took place over the disused Chilbolton Airfield, near the site of the radio telescopes. One spin recovery proved to be particularly nerve shattering. We entered the spin by the usual stall technique which turned out to be a leisurely fairly flat spin but when I applied the spin recovery controls the aircraft went into a very rapid nose down spin before recovering which frightened the life out of me. At the other end of the scale, there were occasions when the aircraft simply refused to enter a spin and the exercise had to be abandoned much to my relief.

On our first cross country flight, we were returning to Thruxton and flying in the vicinity of Lasham Airfield when we quite suddenly found ourselves surrounded by gliders which seemed to be hovering to all appearances like raptors in search of prey below. On my first solo outing, I was doing circuits of the airfield as instructed when I was warned of an approaching thunderstorm by air traffic control and set off on my final circuit. The storm came very quickly and my final circuit and approach was very dark and very bumpy.

My first solo cross country was also something of a disaster as after flying from Thruxton to Portsmouth, across the Solent to Sandown on the Isle of Wight, back across the Solent and along the coast to Shoreham, where I was to do my first landing. Unfortunately, the main grass runway was roped off which was not mentioned by air traffic control so I wondered what to do. I opted to land alongside the main runway but this proved to be very bumpy and I made a heavy landing and the aircraft tipped onto its nose and bent the propeller. The airfield rescue responded immediately and found me shaken but uninjured. Thruxton was contacted and Barry collected me in another aircraft.

After around 18 months, I completed my training and obtained a private pilot's licence (PPL) but for reasons which will become clear later, did not continue with flying.

Richmond

During this period, I attended another Electronics for a mechanical engineers' course, this time at Richmond College. I thought it would be worth doing it again as the technology concerned had moved on quite a bit since 1966.

I was accompanied by Dick Story and Dennis Sharp, two other engineers from Technical services and we stayed in a B&B in a tree lined avenue near to

Wimbledon Tennis Courts. Both Dick and Dennis proved to be heavy drinkers and we ended up in the evenings in a very pleasant pub alongside the Thames at Thames Ditton. On the first night, my attempt to keep up with them resulted in me being quite sick that night and after that gave up the unequal struggle and drank at my own pace.

The classroom where the lectures took place overlooked Richmond Park and the River Thames affording a view like one that Turner would have painted 150 or so years before.

More Hi FI

At about this time, I started tinkering with Hi Fi again. On one of our visits to my parents', I spotted a slice of an elm tree trunk in Dad's timber hoard and thought it would make a great turntable base. I then assembled the components to make a turntable. The tonearm I was going to use had a problem with its wiring so I decided to seek help from Naim Audio in Salisbury.

When I arrived there, I was met by a man who I immediately recognised from Hi Fi magazine articles as Julian Vereker, the MD and founder of the company. When I showed him the tonearm, he dismissed it as rubbish, opened a desk draw and handed me an old Hadcock tonearm, saying use this instead.

When I asked him the price, he said, "Nothing, try it and be impressed."

Naim later produced their own tonearm which used the same type of unipivot bearing as the Hadcock and earlier Decca London tonearm. Needless to say, I purchased one.

The Naim 'factory' then was a cottage in Salt Lane, which later became a fish restraint. The staff at that time consisted of Julian, a lady who was his business partner and accountant, a chap called Ian Kenny who did the final testing of amplifiers and a staff of half a dozen or so, mostly girls, who did the wiring, assembly and packaging.

I later built an amplifier from an article in Hi Fi News, obtaining the electronic components from Radio Spares. I configured my version as separate pre and power amplifiers and designed and built my own cases from 2"x1" aluminium alloy channel and 10swg aluminium alloy plate, in my version of the Naim 'bolt together' boxes of their early amplifiers.

Detached Trials

During my period at D Squadron the Lynx release to service trials were in full swing and these included quite a lot of detached trials. These covered three areas. The first was snow, icing and low temperature trials which were carried out in Canada, at Ottawa, Cold Lake and Fort Churchill. The former was somewhat unusual in being both a military and civil airfield.

The second was hot and high trials which were usually carried out at El Centro in California, Alamoso in Arizona and Leadville in Colorado. One unusual incident occurred on return from one of these detachments when an infestation of Black Widow spiders was discovered and a specialist team of pest controllers had to be called in.

The third was ship trials to establish ship's operating limitations. My only direct involvement in ships trials was a visit to Devonport Naval Dockyard to view a Type 23 frigate's facilities for a forthcoming trial. We flew down to Devonport in a Wessex helicopter and landed on the sports field. I found the frigate to be very claustrophobic including the tiny wardroom where we had lunch.

On one of these trials, there having been complaints that the same 'elite teams' were always sent on these detachments, the principal supervisor, Stan Snowdon, and I decided to send a 'worst team' who had never done a detachment. This caused some consternation, especially among flight test engineers and Lt Cdr Simon Thornuel, who was the team leader and test pilot. However, the team actually performed magnificently and the trial was a success, in spite of the fact that the only Lynx available for this trial was XX 510, a development batch aircraft which was non-standard and had a reputation for unreliability.

The only mishap during the trial also dragged me into it. The auto flight control system (AFCS) computer failed on a weekend and Simon rang me at home to organise a replacement, having failed to raise any of the people on his contact list. Unfortunately for him, Carole answered the phone and thinking it was a wind up by a practical joking neighbour, kept responding to Simon's entreaties with, "It's you isn't it, Roy?" Much to Simon's increasing frustration. Carole eventually caught on and handed the phone to me. Simon then gave me the Westlands part number of the nonstandard computer and a location in RNAS

Portland to despatch it to. I then went into Boscombe Down, located the computer in the main stores and arranged for its despatch to Portland.

Another Handshake with Royalty

In the late autumn of 1978 or 1979, Prince Charles visited Boscombe Down. The main reason for the visit was to hand out the prizes at the annual ETPS McKenna Dinner, but it also included visits to key facilities, including D Squadron. After being introduced, I started to talk about our current work on the Lynx helicopter, which was met with a puzzled expression on his face.

His appearance could only be described as immaculate and as one of my supervisors said, "He looked as if he was daily bathed in asses' milk."

I am not sure which bit of the Old Testament this quote came from but I think it probably refers to a lady rather than a man.

A photograph was taken of me shaking hands with the Prince and I gave Mum and Dad a copy which took pride of place on their lounge wall for the rest of their lives. This was a very youthful Prince Charles, long before the advent of the trauma of Diana and Camilla. I no longer have a copy myself; it is among the things which have disappeared without me consciously disposing of them. The only item in this category I have very deep regrets about is Carole's letters.

I did not like to tell Mum and Dad but I am more of a republican than a royalist, being of the opinion that it all went wrong when Oliver Cromwell died. I am not sure that a constitutional monarchy really works as operated in the UK. Most of the recent monarchs have done their best with the very limited and purely ceremonial role imposed on them. My impression is that the only one to try and break out of this and actually do something useful was Prince Albert.

Our Troublesome Supervisor

When we lost one of our mechanical supervisors on promotion, the Aircraft Servicing Manager insisted that we have this individual as a replacement, much against the wishes of me and my Principal Supervisor, Stan Snowdon. By this time, the Aircraft Servicing Manager was Pete Stevenson; the ex-Engineer i/c B Squadron who was approaching retirement and had been temporary promoted to fill the post. We did not get on and were frequently at loggerheads during the rest of my period in D Squadron, probably as much my fault as his.

The problem for us was that our replacement came with a reputation. His problem was heavy drinking, which rendered him irrational, aggressive and abusive. In one previous incident, he was banned from the Sargent's Mess after smashing up furniture during a fracas. During his period with us he was involved in an even more serious incident. He was leaving a pub in his Triumph GT6 when he was overtaken by a police car with flashing blue lights. When the police car pulled in for him to stop, instead of stopping he accelerated away resulting in a chase along the Woodford Valley road before he finally stopped. When the policeman opened the door of his car, he became aggressive and the police had great difficulty in getting him under control, resulting in him spending the night in a police cell. This resulted in a court case and coverage in the local press.

As a supervisor, he was very limited technically but was good at organising people and tasks and getting things done. He was in fact the leader of the servicing team on the successful Lynx ship trial covered above. He was also the organiser and leader of a team of Boscombe Down volunteers who provided the ground movements of aircraft at the annual International Air Tattoo (IAT). In this capacity, he was highly regarded by the organisers of this event, including 'celebrities' like WW2 air ace Douglas Bader and renowned test pilot Neville Duke.

He was not very good at remembering people's names and tended to give them the all-purpose name of 'Shag'. He was also no respecter of persons and anyone, however exalted, could end up being called 'Shag'.

Our Promiscuous Supervisor

One of our more senior supervisors had a well-deserved reputation for being promiscuous, not in any sort of predatory way but just having 'lady friends' on the side. I saw him with his wife on a couple of occasions at the annual squadron Christmas parties and they seemed to get on fine and be very happy together. I assume that she knew about his 'affairs' and simply tolerated them. He was usually the servicing team leader of the Ottawa detachments and had a regular lady friend there. While in Ottawa, he did not socialise but stayed in her house watching TV in his slippers. They were apparently like an old married couple.

Our Harassed Electrical Supervisor

One of our two electrical supervisors was not a natural leader in spite of being very knowledgeable and competent technically. Unfortunately, he was burdened with an insubordinate charge hand and a few unruly craftsmen. They constantly tried to make his life a misery although he was capable of holding his own. This proved to be a bit disruptive and Stan and I frequently had to reign things in. In fact, two of the individuals concerned were later promoted, took up posts elsewhere and proved to be very competent, even outstanding, in these posts.

Carole and Cancer

In December 1980 after a biopsy to investigate an inflamed hip, Carole was diagnosed with a form of cancer called sarcoma. Her first treatment was radiotherapy which was carried out over the period from December until March at Southampton General Hospital. She was then switched to chemotherapy, then a fairly new treatment which was carried out in the six-month period until September at the old Salisbury Hospital in Fisherton Street under the supervision of a specialist from Southampton Hospital.

This treatment rendered Carole severely disabled meaning that I was spending a lot of time and effort on being her carer to the detriment of the job. To this end the Superintendent of Technical Services, by then, Ken Reid decided to switch me to a less demanding post as Engineer i/c Mechanical Workshops.

Lynx XX 510, undergoing ships operating limitations trials of a type 23 frigate. Another composite from two photographs. I used artistic licence to 'liven things up a bit'.

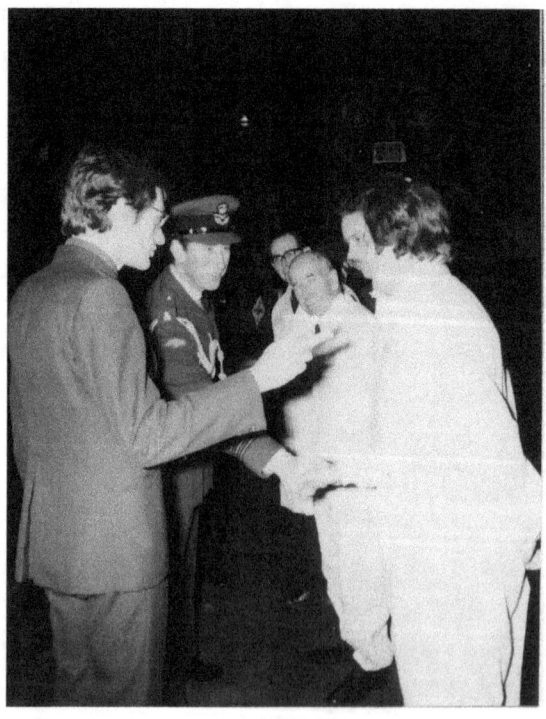

Introducing Prince Charles to some of my tradesmen who did not normally wear white overalls.

Chapter 11
Mechanical Workshops

1981-1984
A&AEE Boscombe Down

Engineer i/c Mechanical Workshops
Management of the establishments mechanical design office, mechanical workshops and bay maintenance facilities.
Projects included:
1. Rationalisation of ground support equipment maintenance.
2. Provision of contractor support for design and manufacture.
3. Market testing of design facilities.
Training and education included:
1. Tornado F2 Systems ------------EASAMS
2. Promotion Boards Seminar – MoD

Mechanical Workshops

As well as the Mechanical Workshops, my new domain included the Mechanical Design Office, the Ground Support Equipment Servicing Bay and Motor Transport Servicing Bay. This grouping in the aircraft orientated Technical Services Department was seen as something of a backwater but things would liven up when my boss, the Specialist Services Manager, retired and was replaced by a somewhat fiery antagonistic individual called George Smith.

The Falklands War

During the period covering the end of my D squadron and beginning of my MWS periods the Falklands War was raging and Boscombe Down became involved in a number of tasks in support of the war effort. D Squadron were involved in modifications to Lynx and Sea King helicopters and in developing parachute systems for airborne forces based on emerging sporting parachute technology. However, the most significant contribution by the establishment was the development of a bomb aiming system for the Vulcan bomber by the Mechanical Design Office and Mechanical Workshops. Boscombe Down was also involved in two modifications to the Nimrod, the installation of flight refuelling and AIM 9 Sidewinder missiles.

Carole Dies

After the chemotherapy and a period of apparent recovery, Carole collapsed with acute chest pains and was hospitalised in late June 1982. An x-Ray revealed that the sarcoma had returned, this time on her lungs and I was advised that nothing more could be done for her except to keep her as comfortable as possible for the short time she had left. This involved very strong pain killers, anti-depressants and other medication to counteract side effects, and quite a lot of time and effort on my part acting as her career. After deteriorating quite rapidly for about a month, she died on the 5th of August, three days before her fortieth birthday.

The Portraits

During the following October I was laid low with a bad dose of flu and had to take sick leave. I had a high fever and, in this state, produced six pencil portraits of Carole from photographs. Although I had produced ink and watercolour landscapes since childhood, I had never attempted pictures of people. Was I attempting to bring her back to life? Or perhaps I was just trying to turn my memories of Carole into art (and probably failing). Who knows! Four of the portraits were not bad and I still have them. My feelings on the loss of Carole are encapsulated in the lyrics of an early Bob Dylan song, 'Tomorrow is a long time' about his feelings on the loss of a loved one. There is a very good version of this song by Sandy Denny on her second solo album 'Sandy'.

George Smith

In the months after Carole's death, my morale reached an all-time low with a state of mind that life without Carole was not worth living. In these circumstances, George sort of became my mentor and saw his role as 'snapping me out of it' and getting me motivated again. This also coincided with another round of defence cuts and the first hints towards a move to privatisation.

The first task George gave me was to carry out a 'market testing' review of the Mechanical Design Office. This proved to be something of a non-event as the very small design office with its low and intermittent workload was not a very attractive proposition for any contractor. In fact, the obvious solution and what we opted for was to use our designers to produce design schemes and a contractor to produce assembly and detail drawings. We found a local firm called Wessex Designdraft who were set up to do just this and tasked them when needed. This worked moderately well but was not without its problems.

I decided to look at contracting out for other areas suffering from low staffing levels and together with John Foy the Principal supervisor we started a market research to look into this. The first area we identified as being worth contracting out was metal finishing which in our case consisted mostly of anodising aluminium alloys and zinc plating steels. The first contractor we looked at was based in an idyllic site alongside the River Hamble near Southampton. In spite of the very pleasant location, the premises consisted of a few sheds and facilities were very basic and the control of work struck us as somewhat hit and miss and a bit chaotic. In the end, we chose a contractor in Wallisdown Road, Bournemouth, who served us well.

George had an unusual approach to staff management and surprised me by asking me to write my own annual staff report. I did so and when he read my effort, he said I had undersold myself and changed some of it for the better.

On the subject of staff reports, I was told about one written by one of my predecessors of my first job as Engineer i/c E Squadron. As second reporting officer he was required to write a summary on the individual being reported on. This individual was seen as something of a maverick and my predecessor wrote

141

'this man is like a cow who delivers a gallon of first-class creamy milk and then kicks the bucket over'.

Defence Cuts and Redundancies

The other driver for our investigation of contracting out was another round of defence cuts which also included a prescribed percentage of staff cuts. The decision was made to exclude staff directly involved with the aircraft; thus, the cuts would be cantered on support staff. This meant that my workshops and servicing bays would take the brunt of the cuts. We achieved this with voluntary redundancies except for workshops final year apprentices who were not taken on as craftsmen on the completion of their apprenticeship.

I found it particularly distressing handing out letters to this effect to the individuals concerned. However, I was able to arrange training in numerical control (NC) machine tools at RAE Farnborough and find employment for them all at an engineering company in a trading estate at the village of Downton, south of Salisbury.

This left us with staff shortages in the machine shop and MT servicing bay. We therefore sought a contractor for the former and eventually chose Enham Industries in the village of Enham Alamein, north of Andover. This started as a community set up following WW2 to employ disabled ex-servicemen. I believe Viscount Montgomery, the victor of the battle of El Alamein was involved in setting up this enterprise, hence the name of the community. I believe that furniture making was the main activity when the enterprise was first set up but by this time had expanded into other activities including a machine shop. They proved to be a very good contractor, providing an excellent service and there was a feel good factor in using them.

For the MT serving bay, we contracted out cars, vans, Land Rovers and buses to local companies and kept the specialist vehicles 'in house'.

Irene Mary Vials

My other saviour from my despondency following Carole's death was Irene. Her first husband, Dave, had died of a heart attack a few years earlier, aged just 42. We exchanged letters several times over the Christmas period culminating in an invitation for dinner at her house in Shillingstone near Blandford Forum on 15 January. During my visit her daughter Loren and boyfriend Terry called in.

The dinner was enjoyable and incident free until we came to the sweet. Irene had prepared a damson crumble but when we came to eat it, we found to Irene's horror that she had mistakenly used sloes which were also stored in the freezer. When Irene took her first mouthful, a look of screwed up distaste came over her face and she said, "Don't eat it," before explaining what went wrong.

In spite of this, we got on fine and I enjoyed the evening. This resulted in us getting together at weekends, including two visits to her son in Margate and culminating in a holiday together in Snowdonia the following August. This holiday coincided with a severe heatwave and we spent most of our time looking for shady places, including the fairy Glen at Betws-y-Coed and a trip down a slate mine. These happy events however were tempered by my feelings of guilt for being 'unfaithful' to Carole.

As our relationship developed, I took her to see my parents at Spetchley and they immediately got on like a house on fire, although Irene initially had difficulty in understanding Dad's Worcestershire dialect. This resulted in us resuming our regular visits to Worcestershire at Easter, Summer Holiday and Christmas. Because of a promise I made to Carole these visits also included me visiting Carole's parents at Worcester, initially on my own but after a couple of years I took Irene as well and they readily accepted Irene, got on well with her and thereafter were always pleased to see us.

On one of our early Easter visits, I took Irene to Bredon Hill and Irene walked barefoot from Woollas Hall to the summit because she said that her sandals were hurting her feet.

Grandchildren

During the period 1983–2010 one of our main recreations was having the grandchildren stay with us. We ended up with three each and different combinations of them stayed with us during the Easter and summer school holidays and half term holidays. We visited all the usual seaside resorts and tourist attractions with them but the most frequently visited location was Hengistbury head and Mudeford Sandbanks. We usually parked in the large car park to the west of Hengistbury Head and Irene and the children would travel to Mudeford Sandbanks on the land train while I walked over Hengistbury Head.

At the end of our day at the seaside, we would usually all walk over Hengistbury Head back to the car park.

When I first met Irene, she already had one granddaughter, Laura, who was her son Mark's daughter, who was born in March 1982. Claire, daughter of Loren's followed in September 1985 and finally, another daughter Kate in July 1987.

My first granddaughter, Emily, was born in December 1991, followed by Oliver in February 1995 and Isabel in July 1997. They gave us constant entertainment and amusement as the following tales will show. My trio grew up in Surbiton so were entranced with their early visits to the countryside of Newton Tony. They got very excited by their first glimpses of cows, sheep and tractors and were intrigued by the several avenues of trees in the area, which they called tree tunnels.

Laura

I first encountered Laura in April 1983 when we visited her parents, Irene's son Mark and his partner Christine, who were living in Margate at that time. We visited them again in June 1984 and this time took Laura to the nearby bird park at Blean. While we were wandering around the park, we came across a black baby girl in a pushchair. When Laura spotted the baby, she stood in front of the pushchair, staring intently at the baby for what seemed an age and eventually pointed at her and shouted out 'baby'.

I don't know why but when, on a later occasion, I suggested visiting the Fleet Air Arm Museum at Yeovilton, Laura was very keen. Of the exhibits, she was particularly keen on the Concorde and when we had gone through the aircraft, she said she wanted to fly in it. When I tried to explain that we couldn't, she said yes, we can, you have flown aeroplanes so you can fly it.

Claire

Claire was daughter Loren's first child and soon came across to me as very confident, bossy and opinionated. On one of her early stays with us, we took her to the New Forest, where we were held up by a herd of slow-moving cows. As the cows slowly walked past us Claire suddenly shouted out, "Those cows have got pooey bums."

Later on, Clair started agitating for a pet and I said I would get her an earwig and a woodlouse. Claire instantly responded with, "Those are creatures; I want a proper pet."

She eventually got her way with a Guinea Pig. In spite of Claire's apparent aversion to 'creatures', when I told them that a snail, she and her little sister Kate spotted in the rockery was in fact 'Brian, the snail', both she and Kate seemed quite taken with the idea.

Later, on one evening, Irene asked me to read them a bedtime story as she was busy doing something. I started to read a story but soon got bored with it and started to make it up, probably motivated to speed things up as well. Clair immediately responded with 'That's not what it says; read it properly'.

Kate

Kate, Loren's second child was, by contrast, a quite shy and reserved child who would occasionally come out with profound statements. When we went to East Lulworth to look at a TVR sports car on sale at a garage there, we took Claire and Kate with us. The owner of the garage was more interested in telling us about the wedding of his daughter then selling the TVR, which turned out to be a dud. After that, we went to a hotel in West Lulworth for a meal and during it Irene said, "He looked too young to have a married daughter," to which Kate responded, "I expect he uses Oil of Ulay."

One regular routine with Kate and Clair was bath night with both of them in the bath splashing about with an assortment of plastic toys. I taught them the words to the songs 'Any old iron" and "My old man's a dustman' which we would all sing along together. Another regular routine was them asking me what I did at Boscombe Down to which I always responded with 'I wind up the aeroplanes'. If they ever get around to reading this, they will find out the truth.

Emily

Emily, son James' first child, was a lively energetic girl and a chirpy individual. In Surbiton, where they lived at the time, we used to quite often come across a Citroen 2CV and Emily was quite taken with this strange, rather cute little car and called it the 'Emily Car'.

One day when we were walking through Newton Tony, she spotted a 30mph sign and responded by breaking into a run and shouting out, "Am I doing 30mph yet?"

She also had the strange habit of taking up a pose every time she saw me with a camera. On one of her early visits to Newton Tony, we took her to my parents in Worcestershire and she was surprisingly patient with the three-hour long journey. While there, she was fascinated with the apples on the small tree in the front garden.

In July 1997, Claire, Kate and Emily stayed with us for three or four days and the four-year-old Emily was so excited at having two big girls to play with that she chattered constantly and became even more hyperactive than usual, which soon led to Clair and Kate complaining that Emily was wearing them out.

One very surreal incident happened when we were driving in the vicinity of Hampton Court, taking the children home after a holiday with us when Emily started 'reading' from an imaginary letter she had written to the Pope, outlining a case for her to be made into a saint.

Oliver

Oliver, James' second child, developed a great enthusiasm for Thomas the tank engine as he reached the toddler stage and this coincided with my father making a large wooden toy train. I took this train, painted it in a 'Thomas' colour scheme and gave it to Oliver as a Christmas present. I still have a photograph of Oliver holding 'Thomas' which was as large as him.

Oliver was prone to the occasional tantrum if he did not get his own way and this manifested itself when we took him and Emily for a trip on the Swanage Railway. The excursion started in the village of Corfe Castle where we purchased a bag of 'dinky doughnuts' at the village bakery and sat on the steps of the old village cross to eat them before setting off for Swanage on the steam train.

On arrival at Swanage, we were walking towards the seafront when Oliver spotted some go-carts going around a small circuit and he immediately wanted to go on one. When we explained that he would not be allowed to because he was too small, he threw a huge tantrum. I am afraid Emily and I just wandered off leaving poor Irene to deal with it, which she eventually did by directing his attention to an adjacent play park with a stationary train in it. Oliver immediately forgot about go-carts, dashed to the train and started to play on it. He was soon organising the other children playing on it and acting as guard and driver.

As he grew older, the tantrums stopped and he became quite a calm measured individual.

Isabel

Isabel, the youngest, spent three months living with us as relations between her parents deteriorated. She was a very cheerful chirpy individual and a persistent watcher of a children's TV programme called TV Tots. When Irene played her CD of Robbie Williams on the Hi Fi, she would respond instantly by performing an energetic little dance.

At this time, my car was the Marcos which had a very loud bellow of an exhaust note which Isabel could hear long before I arrived home. As I drove up the steep drive she was jumping up and down with excitement behind the kitchen stable door with the top door wide open. At our family dinners, she would entertain us all with 'Pythonesque' funny stories in a variety of funny voices. As she reached teenage years and beyond, she became much quieter and reserved.

Promotion Boards and Student Engineers

As my confidence and motivation returned, George Smith suggested that I should become a promotion board member and tutor in the MoD student engineer scheme. I agreed to both with some trepidation and was immediately enrolled on the Promotion Boards Seminar. I do not remember any formal training for the student engineer tutor role other than a visit to the training school at RAE Farnborough where students received 12 months of engineering training before commencing university. I started to fulfil both roles almost immediately and soon came to quite enjoy both.

During the rest of my career I was tutor to three MoD student engineers who obtained engineering degrees at UMIST (Manchester University), Bath University and Brunel University in London respectively. I arranged summer vacation placings for them at Rolls Royce Small Engines Division at Leavsden, WRE Fort Halstead, RSRE Malvern, CDE Porton Down and RAE Farnborough as well as Boscombe Down.

Demands for my services as a promotion board member gradually increased over the years until I had to start declining on some of them. The venues for these included AWE Aldermaston. RSRE Malvern, the Royal Signals Regiment at Blandford Camp, RAE Bedford and MoD offices in Bath, Basingstoke and London.

Three of these were particularly memorable for me. The first was a review of graduated Mod student engineers in one of the London offices. One of the candidates was a graduate of Cambridge and his response to all the boards questions was a lecture on the topic, an astonishing tour de force, the like of which I had never seen before or since. I wonder what happened to him.

The second was at the Empire Hotel in the centre of Bath, which was then owned by the MoD, presumably commandeered during the war and used by MoD Navy. I expect that the building is now luxury apartments. The conference room where the interviews took place was on one of the upper floors of the multi floor building and had stunning views over the sunlit sandstone buildings and terraces of Georgian Bath. I was so distracted by the view that I had to be nudged into life by one of the other board members when it was my turn to interview the first candidate.

The third was at AWE Aldermaston and we were interviewing tracers from the mechanical design office for promotion to design office assistant, a sort of junior draughtsman. As all the candidates were ladies, some of them quite young I was at something of a loss as to what to question them about. However, I was pleasantly surprised to find that some of them turned out to be very knowledgeable on engineering and I was able to broaden the scope of my questions quite considerably.

Moving On

In the mid-1980s, Technical Services Division and Flying Division decided to reorganise their squadron set ups in response to a change in the balance of aircraft types at the establishment and the formation of a RAF Tornado Operational Evaluation Unit which needed a hangar to accommodate it. This resulted in reducing the squadron engineers from four to two. The Engineer i/c D Squadron remained unchanged except for a name change to Engineer I/c Rotary Wing Test. The other three squadron Engineers were replaced by a single

Engineer i/c Fixed Wing Test who was also responsible for the ETPS fixed wing aircraft. I was offered this post and accepted it.

One of my Carole pencil portraits.

Irene at the Fairy Glen at Betws-y-Coed, one of our shady places.

Being fed by Granddaughter Emily.

Chapter 12
Fast Jets and Heavies

1984–1992
A& AEE Boscombe Down

Engineer i/c Fixed Wing Test & ETPS
Management of the maintenance and preparation for trials of all the establishments fixed wing aircraft, including those of ETPS.
Projects included:
1. Review of Comet structural maintenance programme.
2. Flight test instrumentation (FTI) fits to ETPS Tornado and Jaguars.
3. Support of release to service trials of Harrier GR5 & 7, Sea Harrier FRS2, Tucano, Tristar Tanker and E3D Sentry.
4, Development and implementation of a performance indicator monitored productivity scheme.
5. Asset/facility study of the new DGT&E organisation.
6. Board of enquiry on the Hawk 200 fatal accident at Dunsfold.
7. Project management of the conversion of the ETPS "ASTRA" Hawk.
Training and education included;
1. Managers Safety Course – ROSPA
2. MoD Productivity Scheme Audit – MoD/Treasury
3. Staff Reporting – MoD

Release to service trials

During my period in this post the release to service trials were carried out on the following aircraft:

1. Sea Harrier FRS 2 – The royal navy version of the Harrier which had the addition of an airborne interception radar in the nose.

2. Harrier GR5 & 7 – These were the UK versions of a joint McDonnell-Douglass/BAE update of the Harrier which featured a composite structure front fuselage and main plane. The US version was called the AV8B.

3. Embraer Tucano – The RAF basic trainer to replace the Jet Provost.

4. Lockheed Tristar – A tanker version for the RAF.

5. Boeing E3D Sentry – The RAF version of the Airborne Warning and Control System (AWACS) aircraft.

The Village Pantomime

From the summer of 1983, Irene became a regular weekend visitor to my home and in June 1984, after the wedding of her daughter, Loren she moved in with me. My abiding memory of Loren's wedding is the wedding dress Irene made for her. It was stunning and although I am no expert on fashion, it is the best wedding dress I have ever seen.

The village started to hold a pantomime for Christmas in 1983 and I was asked to do the scenery the following year. We decided to use four door sized plywood panels suspended from a frame at the back of the stage for this. Each pantomime had six changes of scenery so we had three sets of panels for me to paint scenes on both sides. Over the years I have produced works of mostly scenery and buildings on A4 and A3 sized cards in pencil, ink, water colour and acrylics. This time my media was a mixture of emulsion, primer, undercoat and gloss paints in half full or less tins provided by a local painter and decorator. I then set to work painting the scenes assisted by Irene.

For the pantomime performances, I directed a team of some of the actors in changing the panels between scenes and these activities tended to be somewhat chaotic and noisy. I did this for three more successive pantomimes before the decision was made to do away with background scenery and rely on props.

Maurice, The Film

During one autumn between pantomimes Newton Tony was invaded by a film crew for a Merchant Ivory film 'Maurice' based on the EM Forster novel of the same name. The film starred Hugh Grant and a host of well-established

British character actors. The director found out that there was a pantomime group in the village and we were approached to provide 'extras. Irene's father Len was staying with us so he was roped in as well. I took part in the cricket match scene and Irene, Len and I took part in the wedding scene. Len, with his mop of wiry, curly hair and long sideburns had a very 'Dickensian' appearance and the director wanted to retain his services but Len declined the offer.

About six months later we were invited to a preview of the film at the cinema in Salisbury and as Len was staying with us again, he came along as well. The version we were shown was an early largely uncut version which lasted a marathon nearly three hours, a lot longer than the final version we later saw on DVD. Len fell asleep after about fifteen minutes, which was just as well as the film was about homosexual relationships.

Maurice, The Landlord

During the pantomime period the cast would, after a rehearsal or performance, retire to the village pub the Malet Arms, to wind down and partake in alcoholic refreshments. The landlord at this time was a retired policeman called Maurice. He was a very loud, erasable individual and people would come for miles to be insulted by him.

On one occasion, a young courting couple were sitting closely coupled in a corner of the bar and when Maurice decided that they were getting too affectionate, he shouted out, "Put a stop to that or I will throw a bucket of cold water over you."

On another occasion, we were sitting in the small snug bar at the side of the main bar with another couple and while the man was ordering another round of drinks from Maurice through the hatch into the main bar, the lady asked him politely if he could put some more logs on the fire as she was feeling cold. Maurice responded by coming around, piling the fire high with logs and stoking it up to a roaring blaze. On departing the room, he turned around and said, "Roast you buggers."

Another trick Maurice would deploy was to let his very large pet goose behind the bar. The goose would appear jumping up and down and honking at deafening volume. If a customer dared to approach the bar, the goose would lunge across the bar at him.

Arts and Crafts

Someone discovered a set of photographs of the village taken in the early 1900s, had copies made for sale in the village shop and a jumble sale in the village hall. I purchased a set and produced four A3 sized ink pictures from them and had copies made at A3 size and smaller versions as notelets.

At the same time, Irene became interested in traditional dolls and purchased some kits which comprised ceramic heads and limbs and plans for making fabric and filling torsos and clothing. She made about a dozen of these and gave some away to her grandchildren. Irene designed the doll's clothes herself and the results were superb. The only snag was that similar dolls were for sale in gift shops for a lot less than Irene paid for the kits and materials.

At the next village jumble sale, we had a stall selling my pictures and Irene's dolls. I am sorry to say that my pictures sold like hotcakes but Irene's dolls did not sell for the reason stated above.

Uncle Perse

Uncle Perse was Mum's sister's husband, who at the time of my re-acquaintance with him had moved from the council house in Hurstbourne Tarrant to a flat in Andover, when the former was demolished to make way for a village by-pass. By this time, Aunty Rene had developed dementia and gave Uncle Perse a lot of problems by wandering off to see her deceased brothers and getting lost in the process. She finally died at about the same time as Carole.

Irene and I would visit him about once a month and Irene would bake a cake for him which he was very partial to. During these visits he would regale us with tales of his childhood and youth in and around Hurstbourne Tarrant and his 'war memoirs', in his very broad Hampshire dialect. After a while, we were getting mostly repeats of stories already herd.

My favourite 'war memoir' was the VD lecture story. In this, the Sargent who was delivering the lecture, invited questions at the end of the lecture. In response to this, one of the recruits held up his hand and asked, "Can you catch off toilet seats?"

The Sargent responded to this with, "No son, only officers and Vickers catch it off toilet seats."

When my son was married, Uncle Perse was invited and during the service, he started carrying on in a fairly loud voice with comments like, "This is the wrong tune for this hymn," and "why doesn't he speak up, I can't hear him." This was met with a mixture of consternation and amusement from nearby worshipers. Then at the reception he repeatedly said how nice the wine was in a loud voice which led to waiters constantly topping up his glass. After a while, he said, "I feel queer," tried to get up and fell over. This resulted in me having to take him home. Luckily our arrival at the flat was spotted by a neighbour who happened to be a retired hospital matron and I explained the situation to her and she said she would see that he was all right.

A few months later he was admitted to Winchester Hospital for observation after more falls. When we visited him, he was in a small ward with four old ladies who were on the receiving end of his childhood stories and 'war memoirs' which they enjoyed very much. On our next visit, we learned that he had been moved to a single room, but the ladies complained so much that they moved him back again.

West Freugh and Llanbedr

These were RAE outstations where the facilities including operation and maintenance of the aircraft were contracted out. At this time, the contractors were a partnership of SERCO and Flight Refuelling with the latter being responsible for the aircraft. The aircraft at West Freugh near Stranraer in Dumfries and Galloway were Buccaneers used for bombing training and trials on the establishment ranges. At Llanbedr, between Harlech and Barmouth in Cardigan Bay, the aircraft were Australian built Jindivik drones which carried out target towing duties over the Cardigan Bay ranges. Not long after I became Eng. i/c FWT the responsibility for monitoring the aircraft contractor was switched from RAE to A&AEE and me.

On my first visit to West Freugh, I decided to take Irene with me and extend our visit as a holiday break, a thing I repeated 12 months later. A feature of this visit was Irene being accosted by the 'village idiot'. The first of these was in a pub in Dumfries, where we had stopped for refreshments en-route to West Freugh. The pub consisted of a long downhill corridor with seats and tables down one side and the bar at the bottom. Irene had seated herself at a table near the entrance while I went down to buy drinks and crisps, when a man came up to her, thrust a banknote into her hand and said, "Get me a drink."

Fortunately for Irene, the barman spotted this and immediately grabbed the individual and unceremoniously turfed him out of the pub. Having done this, he returned, apologised to Irene and explained that the individual was a drunk who had been banned from the pub.

We stayed at a small hotel alongside the beach at Sandhead which was called the Tighe na Mara which we were told was Gaelic for 'by the sea'. I am not sure that I have the spelling right though. That night we had encounter number two when, after a very large and pleasant dinner, we were having a final drink before going to bed. This time a rather strange man who was wearing a Stetson hat with 'Australian style' corks hanging from it, offered to buy Irene and I a drink. We accepted and when we were finishing these drinks, he offered to buy us another round. Fortunately, this was spotted by the landlord who told him to desist. Apparently, a strict closing time was observed but residents were exempt from this rule so this individual was trying to extend his drinking time.

On our second visit together, the Tighe na Mara was full so we stayed at the Torrs Warren Hotel, which was on a hillside overlooking Luce Bay and this time Irene was involved in a 'Faulty Towers' incident with the landlord while I was attending the quarterly review at the establishment. This disturbed her a bit, but she could see the funny side too.

I was pleasantly surprised to find my short time boss from my RAE Bedford period Ted Payne attending as the Farnborough representative. As the quarterly review finished early, Ted and I picked Irene up from the Hotel and we visited Logan Tropical Gardens and the quaint coastal village of Port Logan.

I also renewed my acquaintance with Ted Whitley another boss from Bedford days, who was in charge of the Flight Refuelling aircraft servicing team having retired from his Aircraft Servicing Manager post at RAE Bedford. On our second visit to West Freugh, Irene and I had a meal with him and his wife Netta at his home near Stranraer. Netta was not very happy living at a location 50 miles from the nearest M&S store.

On other visits to West Freugh, I availed myself of Boscombe Downs Piper Navajo Chieftain 'air taxi'. Two of the return trips stand out in my memory. On the first, we flew over the Solway Firth, the Lake District, Morecambe Bay, Blackpool seafront, the Long Mynd, the Malvern Hills and Severn Estuary in a rare treat of a scenic route. On the second, we hit severe turbulence over the

Pennines with me being continuously thrown up out of my seat and banging my head on the luggage rack, in spite of the seat belt being pulled as tight as I could get it.

I only visited Llanbedr once and after the quarterly review I was shown around their facilities and aircraft. As well as the Jindivik drones they had a Hawk which had been converted to combined drone control by a ground pilot and normal control by a master-pilot. This was used to train the ground pilots in drone operation. In previous years, quite a number of Meteors and four Sea Vixens had been converted to drone operation.

Comet 4C XS 235

This long-term aircraft was used by Navigation and Radio Division as a flying laboratory aircraft for testing and recording the performance of navigation and communications systems. This was the last Comet built and was given the name 'Canopus'. With the age of this aircraft, we were concerned that the original structural integrity checks in the maintenance schedules were no longer adequate. In reviewing this and compiling a new schedule, I received great assistance from BAE Woodford and the Nimrod team at RAF Kinloss.

The most serious problem this exercise revealed was the main undercarriage which BAE could no longer overhaul because key spares were no longer available. The solution BAE came up with was to replace it with the Nimrod main undercarriage. As the change of undercarriage was beyond our capabilities and facilities, I approached RAF Kinloss and they agreed to take on the task.

At the beginning of January, immediately after the Christmas break, we departed Boscombe Down bathed in freezing fog, on the Comet's delivery flight to Kinloss. When we landed at Kinloss, it was warm and sunny, with people walking about in shirtsleeves, an incredible temperature inversion. The change to the Nimrod undercarriage was not without its problems but the Kinloss team did a magnificent job helped by a working party from BAE Woodford.

My Second Wedding

Irene and I were married on 4 January 1986 at Boscombe village church with the reception at Boscombe Down Officers Mess. The proceedings were videoed by Irene's first husband's brother, John and we still have a very long unedited video of our wedding.

For our honeymoon, we decided to spend two nights in a hotel in central London, near to the Houses of Parliament and Westminster Abbey. The main thing I remember about the hotel was the breakfast with a vast Aladdin's Cave selection of food on display. For our evening meals, we went to an Italian and an Indian restraint, both in Soho. For our daytime activities, we limited ourselves to walks across St James and Hyde Park and the shops in Oxford and Regent Streets, all of this in very cold weather, including flurries of snow. I wonder now why we chose to get married in January.

Mac and Viv and France

One of the guests at our wedding was Mac and Viv. Mac had been a fellow apprentice at the College of Electronics in the 1950s and we were reacquainted when I came to Boscombe Down in 1977, when he was holding down the post of Engineer i/c Air Traffic Control. Carole and I and Irene and I became long term friends with Mac and his wife Vivienne and we frequently swapped hosting dinner parties.

Between 1985 and 2005 we made six trips to France and two to Belgium and Holland with them. Areas visited included the Alpine Region, the Auvergne, the Alsace, Provence, the Pyrenees, Brittany and Burgundy, as well as Bruges/Ghent, Ostend and the Keukenhof Gardens near Rotterdam.

These holidays were very enjoyable and helped considerably by Mac and Viv's reasonable command of French. In contrast, my command of French was meagre and Irene's non-existent. In fact, during our early visits Irene was prone to break into the German she remembered from her time as a military wife in Germany in the 1960s, at supermarket checkouts. Perhaps, I will, one day write a book about France; or perhaps I will just let the photographs tell the story.

Irene's Jobs

When Irene first moved in, she decided that she needed something to occupy her while I was at work and started to look out for volunteer jobs. Her first job was at Amesbury Care Home, where after training in manicure and pedicure by the Red Cross she applied her new skills to the inmates, who also enjoyed and appreciated her company.

This only provided employment for one morning per week so she looked out for further employment and found a job as a receptionist at the Marriage Guidance Council clinic in Salisbury. She then started as a volunteer on Wednesday evenings at the Burns Unit at Salisbury Hospital.

Although she enjoyed these jobs and got a lot of job satisfaction from them, she felt that she needed something nearer to full time employment so applied for a job advertised in the local free newspaper at Porton Down Public Health Laboratories, this time, a paid job. The job she was offered was as a messenger, which involved distributing internal mail and samples to offices and laboratories. She accepted the job but only stayed there for just over a week before resigning as she found the atmosphere there somewhat intimidating.

Finally, a job became available as a playground supervisor and dinner lady at the village primary school for which she applied and was successful. She stayed in this job until my retirement and enjoyed it most of the time. She had a good rapport with the children who seemed to like her. On our evening walks around the village and local footpaths, it was not uncommon for a child's voice to shout out from a garden, "Hello, Mrs Bowles."

The Hawk 200 Fatal Crash

The Hawk 200 was a single seat fighter version of the Hawk advanced trainer BAE had developed for hopeful sale to foreign air forces. The prototype demonstrator aircraft, ZG200, was involved in a fatal crash at BAE Dunsfold in June 1986 and I was asked to be the engineering member of the board of enquiry into the accident. The other two members of the board were pilots, a wing commander as leader and a squadron leader as the aircrew member.

On the first day, we visited the crash site, which was basically a very large hole in the ground and saw a film of the flight which was not very revealing except to show that the crash was at the bottom of a loop manoeuvre. At the end of the first day, we had a session to plan our investigation. I was allocated the task of looking at the servicing records while the other two would look at the pilot aspects including his medical records. We expected this to take two or three days and would meet at the end of each day to discuss progress.

In my enquiry into the aircraft servicing, I found that BAE Dunsfold operated a system which mirrored the system the RAF used for their Hawks including a Form 700 'diary' to record all the maintenance activity, including signatures by authorised personnel. I found these records to be well kept and there were no deferred defects or flying limitations recorded at the time of the accident. The BAE Dunsfold staff were extremely helpful and open during this investigation.

The other two members found that the pilot, John Hawkins, who was known to one and all as 'Jim' after Robert Louis Stevenson's 'Treasure Island' character, was a very experienced test pilot, with the usual career path of RAF/ETPS/MoD (PE)/BAE. His most recent annual medical examination had revealed no health issues.

Having completed this part of the enquiry and while we waited for data from the recovered accident data recorder (ADR), we decided to seek the advice of areas of expertise who could provide advice or practical help.

Our first visit was to the RAF Institute of Aviation Medicine, who gave us a thorough briefing on the medical aspects of 'aerobatics and the current state of equipment's designed to combat ill effects of this such as anti 'G' suits. Having viewed the film, they expressed the view that the manoeuvres seen should not have posed any medical problems. I believe though that aeromedical opinion has since changed somewhat and it is now thought that G induced loss of consciousness (G-loc) could be a contributing factor to an accident like this.

Our next port of call was the Air Accident Investigation Branch, also located at Farnborough. Here we were told that because of the severity of the impact, very little of what would be recovered would reveal much about the cause of the accident and the main hope would be any data recovered from the ADR.

A visit to the ADR section at Boscombe Down revealed that the badly mangled ADR was still in Mechanical Workshops being 'cut open', but in spite of its state, hopes were high of getting useful data out of it.

During these investigations we were told about a facility called the Combat Simulator in Engineering Physics Department at Farnborough, which it was suggested could provide useful information from the ADR data. A visit to Engineering Physics confirmed this and we were told that they could provide a complete visualisation of the flight as revealed by the ADR data.

Having retrieved the ADR data in the format Engineering Physics needed, we handed the data over to them and while we waited, we listened to the cockpit voice recorder output. There was no talking, just heavy breathing and the sound of heavy exertion just before the recording ceased. I found this quite distressing.

The Combat Simulator did indeed give us the simulation of the flight and revealed that the cause of the crash was that the final loop manoeuvre was simply carried out too low. In other words, the cause was 'pilot error', notwithstanding my earlier observations on G-loc. In fact, we later saw a film of a 'Jim' Hawkins display at a Farnborough Air Display the year before and, in the loop, manoeuvre the bottom of the loop was so low that a cloud of dust was raised.

In fact, the same thing happened again at the 2017 Shoreham Air Display and this time the crash of a vintage Hunter aircraft resulted in the deaths of 11 innocent bystanders. I wonder if there is a common factor which caused the pilot to start the manoeuvre at too low an altitude.

The Astra Hawk

Among the ETPS fleet was the Variable Stability Bassett, a twin-engine light transport aircraft which had been converted to be programmed to provide variable stability by having changeable control laws of the auto stabiliser. This was now 'old technology' with the control laws changed by a bank of potentiometers, so a modern version utilising active control technology was required.

To this end, Cranfield, who were already working on an active control system for the RAE Bedford two seat Harrier, were approached. The mastermind behind Cranfield's active control technology was David Williams, who was probably the leading authority on the subject in the UK at the time and later developed the active suspension system used by the Formula One Lotus.

Perhaps a little about active control would not be amiss at this stage. The handling of an aircraft depends largely on the position of its centre of gravity (CofG). A forward CofG results in a very stable aircraft which tends to carry on flying straight ahead with any manoeuvring requiring a lot of effort by the pilot. An aft CofG results in a very agile aircraft but one which is very difficult to control due to over reaction to control inputs. A combat aircraft needs to be very agile but also it must respond precisely to pilots control inputs.

Active control achieves this by inputting the pilot's controls to a computer instead of directly to the control surfaces. There are also inputs to the computer from sensors such as rate gyros, accelerometers, strain gauges and air data sensors. The computer software then enables the pilot input to be immediately actioned while avoiding catastrophic aircraft manoeuvres such as stalling or spinning or structural damage to the airframe.

The Cranfield ASTRA system, (Advanced Stability Training and Research Aircraft), takes this a stage further to create a 'versatile' active control system where the control laws in the computer can be changed to give, for example, bad handling characteristics.

My involvement was acting as project manager for the conversion and attending the progress meetings at Cranfield. These were as usual fairly cumbersome affairs, attended by the active participants, myself, ETPS and Cranfield, plus participants from the MoD (PE) project office and quality directorate, RAE Structures department and BAE. David Williams was obviously as bored of these proceedings as I was and usually appeared to be asleep. However, when asked a direct question he answered promptly and precisely with a hint of distain. I quite liked this and became something of a fan of his. We usually travelled to the meetings by air in one of the ETPS fleet, ranging from the Andover transport aircraft to the Lynx helicopter which was a cause of caustic comments from the other participants.

The BAE presence, as the design authority for the Hawk, was to advise if the Cranfield modifications would affect the clearance of the aircraft and if so, what steps would need to be taken to clear the modified aircraft. Unfortunately, the BAE team changed three times. It started out with a team from Kingston who seemed disinterested and there under sufferance. After about two meetings, this changed to a team from Brough, which was a change for the better. This team were very enthusiastic and perhaps like me could see the possibilities for future marks of Hawk. Finally, the leadership of the BAE team was taken over by Warton but still with a Brough presence. This changed their attitude to that of wanting nothing to do with the modified aircraft, and handing over its design authority to Cranfield/A&AEE. They had a point perhaps, but the Hawk project office were against this and insisted in a continuing BAE presence.

The RAE Structures Department presence at the meetings was to advise on structural integrity aspects of the aircraft modifications. I sought their advice in dealing with the BAE concerns about structural integrity dangers which were countered by Cranfield's counter argument that structural integrity would be improved by the active control system and the expected low utilisation of the aircraft compared with that of normal RAF Hawks. Their response was somewhat noncommittal and they tended to side with BAE. To be fair to them, there were growing concerns about the fatigue life of the Hawk airframes and main planes.

I then initiated direct negotiations with a still reluctant BAE and with a lot of arm twisting, and probably money as well, they agreed to carry out the comprehensive strain gauge fit they used during the development of the Hawk and provide advice on an initial flight test program to establish a datum. This proved to be a very protracted programme and I had moved on long before its conclusion.

The Meteorological Research Flight

This unit, a part of the Meteorological Office, had operated as a lodger unit at RAE Farnborough since its formation and over the years had operated a number of converted laboratory aircraft. The current aircraft was a C130 Hercules, which had been converted to the metrological research laboratory role by Marshalls of Cambridge, who were the nominated UK design authority for the C130.

In 1990, the aircraft was transferred to Boscombe Down but the unit's staff remained at Farnborough. Operating and maintaining this aircraft was no problem for us as we had operated a number of the type over the years. Carrying out modifications to the laboratory fits involved us in a number of visits to Marshalls of Cambridge to obtain information and drawings.

The First Gulf War

Boscombe Down became actively involved in this conflict, which like the Falklands War involved a lot of late evening and weekend working. Most of the tasks involved enhancing navigation and weapon aiming systems on Tornado, Jaguar and Buccaneer aircraft.

The Craft Productivity Scheme

Establishing the pay of civil servants has always been a difficult and contentious issue, with pay usually decided by comparison with equivalents in industry. This has always been complicated by issues such as the perceived generous pensions and job security of civil servants compared with industry norms. Craftsmen's pay was, if anything, even more difficult to establish so The Treasury decided that a new approach was needed and set up a small team to look into productivity schemes.

To this end they first of all organised a seminar which they called 'Audit of MoD Productivity Schemes' and sent invitations to all the MoD employers of significant numbers of craftsmen, including Boscombe Down. As usual when something out of the ordinary turned up, I was asked by the Superintendent to take this on. By this time, our Superintendent was a retired RAF Group Captain who was a very likable individual who had a very positive attitude to the craftsmen. This project had very little engineering content, but I found it quite interesting and ultimately rewarding.

The Treasury team had two people I dealt with. The team leader was a portly rather taciturn individual and his main assistant, Sue Williams, who seemed to do all the real work. For craftsmen like ours who were involved with maintenance and operation, rather than production, a technique called Random, Rated, Activity Sampling (RRAS) was deemed to be the most appropriate. We agreed to initiate this scheme and our trades union representatives were, surprisingly, quite enthusiastic about it. This technique is best explained by its title and the random timings could be stipulated by a random number generator. For example, if an average timing of five minutes was needed, the generator setting of one to ten would be set. The ratings for the activity varied from a minimum for resting or waiting time to a maximum for high physical exertion or heavy concentration.

I was allocated a principal supervisor as team leader, a supervisor and four technicians to do the surveys, which enabled me to deploy two teams to do the sampling. With Sue Williams, I organised training at a nearby army unit at the old RAF Andover airfield site which included carrying out RRAS surveys 'for real'. This done, the surveys began in earnest at Boscombe Down.

To obtain other users view the superintendent and I made a visit to RAF Kemble which was at that time in the hands of the USAF who were carrying out maintenance and modification programmes on A10 Warthog 'tank buster' aircraft using British hanger staff. The day of our visit coincided with Thanksgiving Day so after obtaining their views on RRAS we were invited to a thanksgiving dinner at their officers' mess. The turkey roast dinner was very good with the only unusual thing being the accompanying marshmallows which proved to be surprisingly compatible.

In spite of my concerns, the surveys were very successful and the results showed a surprisingly high level of productivity. To satisfy ourselves that our surveys were not biased we employed a 'guest team' from another unit and their results were no different to those of our teams. As a result, the superintendent and I became very popular with the trade union representatives and industrial relations were the best I have ever experienced.

Changes in MoD and Moves Towards Privatisation

During my earlier career the R&D Establishments and associated ranges were part of the Ministry of Supply but after a brief period as the Ministry of Technology under the Harold Wilson Labour government it became the MoD Procurement Executive for most of my career. That is until 1991 when the research establishments became the Defence Research Agency (DRA). Under agency status the funding changed from an annual 'vote' to operating a trading fund from payments by 'customers'.

Boscombe Down and the ranges were initially excluded from the agency and instead were set up as a separate organisation under a Director General Test and Evaluation (DGT&E). The new Director General decided that a 'doomsday survey' of his organisation was needed and asked Boscombe Down and the ranges to provide members. As usual, I was chosen as the Boscombe Down member and the other two members, representing the ranges, were from Aberporth and Shoeburyness. Having chosen me, the Commandant of Boscombe Down asked me to liaise with him to keep him informed on what we were up to.

We held our first meeting in a conference room of a MoD office block in a dingy, run-down area south of the Thames, which would become the location for our regular progress meetings. Having established exactly what information the Director General wanted, we decided to hold initial meetings at Boscombe Down, Aberporth and Shoeburyness for a conducted tour by the host, to get a feel for the organisation of these establishments. We also decided that I would provide the report on Boscombe Down and the other two would cover the ranges. My task was made easier because over the years Boscombe Down had produced a number of brochures which I could use to describe the task, facilities and organisation of the establishment, leaving me to provide such details as financial arrangements and staffing levels. Two or three months and the same number of meetings later, we presented our report to the Director General but I am not aware of any actions taken as a result of it, mainly because the organisation did not stay in existence long enough.

Promotion

In 1992, I was invited to a promotion review board at one of the London offices and was successful, giving me a 'ticket' to the exalted civil service grade of 'principal' or 'unified grade seven'.

All I had to do now was to find a suitable vacancy to enable me to 'cash my Ticket'. I spotted a vacancy notice for a post at RAF St Athan near Cardiff and applied for it. This resulted in an invitation to an informal interview with the unit commander in chief which I duly attended. At the interview, the commander in chief, a RAF Group Captain, told me that the post was currently held by a RAF Wing Commander and he said he did not wish it to be held by a civilian. With this in mind and not being particularly attracted to the job as described to me, we both agreed that this was not the post for me. Having got that out of the way I was treated to a comprehensive tour of the facilities and a pleasant meal in the officers' mess. At that time, the unit was mainly engaged in a deep servicing and modification programmes on Tornado aircraft, supported by BAe working parties.

My problem was solved when Mac, who was by then the Specialist Services Manager, decided to take early retirement on health grounds with his deteriorating health, due to diabetes and other health issues. As a result, I was offered the vacated post, which I duly accepted.

One of my pantomime scenes.

Irene, dressed for her part in the film 'Maurice'.

Astra Hawk xx 341 over Abbotsbury – Another composite from two photographs. An unlikely location but it makes a reasonable composition.

Comet XS 235 over Maiden Castle – Another 'composite' in an unlikely location, this time a hillfort between Dorchester and Weymouth.

Part 6
Principal Engineer

Chapter 13
Chief Designer

1992–1997
A&AEE/DERA Boscombe Down

Aircraft Installation Manager (Chief Designer).
Management of the support services including, design offices, workshops, servicing bays, avionic rigs and photographic services.
Projects included:
1. Preparation of the A&AEE bid to support the DRA air fleet.
2. Obtaining Def Stan 05–123 Design Approval and ISO 9001accreditation.
3. Management and design certification of the following aircraft installations;
a) A Heli-Tele system converted for ship detection in a Sea King helicopter.
b) Icing trials flight test instrumentation (FTI) and two 900gallon water ballast tanks in a Chinook helicopter.
c) A development Marconi Blue Hawk airborne interception radar in a HS 125.
d) Tornado flying laboratory role fits.
e) Electronic Flight Instrumentation System (EFIS) in a BAC 1–11 and Lynx helicopter.
f) Tucano FTI for ETPS.
g) Specialist equipment fits in the Meteorological Research Flight Hercules.
Training and education included;
1. Business Practices Module 1 – DERA.
2. Managing People and Performance – DERA.
3. Health and Safety for Senior Managers – Cricklade College.
4. RMRP in DTEO – DERA.
5. DTEO 2000 MDP – DERA.

MoD Design Approval and ISO 9001 Accreditation

Almost as soon as I started in my new role as Specialist Services Manager, I was 'rebranded' as Aircraft Installations Manager (Chief Designer). The reason behind this was the decision to seek the status of a MoD Design Approved Organisation and ISO 9001 accreditation.

Our bid was to be assessed by the Engineering branch of the MoD Procurement Executive Director of Flying and the Defence Quality Agency (DQA). My main role in this was a comprehensive review of our design and manufacturing procedures and supporting documentation with the key document being the Certificate of Design.

This proved to be a somewhat tedious but relatively straightforward exercise with the only new procedure being design review and the only new form required was the Certificate of Design. I did some research on these and compiled a suitable procedure and form. However, this still left me with concerns as to when and how the design review would be applied and this was to give me a lot of problems later.

The assessment of our organisation, procedures and documentation proved to be fairly straightforward and we duly received MoD Design Approval and ISO9001 accreditation.

The DRA Air-Fleet

At the same time as we were seeking MoD design approved organisation status, the DRA decided that the expense of running two airfields at Farnborough and Bedford was not a good business proposition, so put the running of their air fleet out to tender. Our bid included the proposed erection of a new hangar and office block to house the aircraft and any transferred staff.

We were successful but I do not think that there were any other serious contenders. The transfer of the aircraft and staff to Boscombe Down turned out to be something of a 'poisoned chalice' for Technical Services Division.

Defence Evaluation and Research Agency

In 1995, DGT&E came to an end and was absorbed into an enlarged agency, renamed the Defence Evaluation and Research Agency (DERA). Initially Boscombe Down and the ranges were set up as a separate organisation called the Defence Test and Evaluation Organisation (DTEO), within DERA. However, after about 18 months and several DTEO seminars of senior staff, including me, to install the DTEO identity including deciding mission statements and suchlike, the idea was scrapped and we became just a part of DERA. Finally, on 1 June 2001 DERA was privatised as a public/private partnership under the trademark of QinetiQ.

Climping and Oundle

I remember very little about the above seminars except for two very pleasant and interesting locations. The first was the Balliffscourt Hotel, south of the village of Climping, west of Littlehampton in Sussex. It was located south of the village not far from the seashore on the site of a Norman chapel. The hotel was originally a manor house owned by Lord Moyne, a member of the Guinness family, but on the death of his widow in 1939 it was converted into a hotel. In 1993, it was purchased by Historic Sussex Hotels and restored and modernised. It was a strange mixture of gothic and modern. While there, I enjoyed walks in the grounds, locality and seashore.

The other was the Talbot Hotel in the centre of the Northamptonshire town of Oundle, which was dominated by a public school. The hotel was an atmospheric coaching inn, built in1626, whose main feature was a large old stately staircase which it was claimed to be from the room in the nearby Fotheringhay Castle where Mary Queen of Scots was executed. There was a large Victorian painting in the lounge of the scene of the Queens execution. There were also claims, I seem to remember of hauntings by Mary Queen of Scots. It was here that I acquired a taste for Laphroaig malt whisky which I was introduced to by Pete Newton, a colleague during late into the night exchanges of yarns in the bar.

The Design Sub Contractor

During this period, I lost some key mechanical design staff including the Mechanical Design Office section leader through retirements and transfers. Recruiting and training up replacements would be time consuming and the workload was building up so I decided to seek a sub-contractor.

To this end, I drew up a specification which included one of our current design tasks for which I asked tenderers to provide a proposal for achieving it. Two of the responses were very thorough and serious. The first was from the ex Folland/HSA/BAE factory at Hamble which now operated as a separate company. The other was a company based at Stanstead Airport which specialised in conversions of airliners for transport and executive roles. In the end, I chose the latter as their pricing was much more competitive.

This proved to be a very successful collaboration which carried out the following tasks over the next three years.

1. Ship detection system in a converted Heli-Tele in a Sea King helicopter.

2. Icing Trials instrumentation and two 900-gallon ballast tanks in a Chinook helicopter.

3. GEC Blue Hawk airborne interception radar in a HS125 aircraft.

4, Flying laboratory conversion of a Tornado aircraft.

5. Electronic Flight Instrumentation System (EFIS) in the ETPS BAC 1–11 aircraft.

6. ETPS instrumentation in a Tucano aircraft.

7. Specialist equipment fits in the Meteorological Research Hercules aircraft.

The electrical design office had no need of sub contract support, being adequately staffed and led by a very competent electrical designer, Noel Cooney. He had been an electrical craftsman in D Squadron when I was the squadron engineer and soon after I moved on, he applied for an electrical design post and was successful. A few years later, when the section leader retired, Noel was the obvious choice to replace him. His work on the ETPS BAC 1–11 EFIS cockpit, which replaced the aircrafts conventional flight instruments with a flat screen display, was outstanding.

The Commandant

The Commandant at this time was Air Commodore 'Reggie' Spiers who was quite a character and the only Commandant that I had dealings with and we did so on amiable terms. My first dealings with him were when I was carrying out the DGT&E 'Domesday Survey' mentioned in the previous chapter.

Our acquaintance continued in my newly promoted post and amongst my charges was Mechanical Workshops, whose engineer was another 'character', Dennis Sharp, who had been involved with 'Reggie' in RN Buccaneer trials earlier in their careers. Dennis was a loud cheerful individual who did not take his job, or life generally very seriously. He was a bit of a raconteur and his version of recent events was always amusing. He was also a very good cartoonist producing cartoons of his version of life at Boscombe Down.

'Reggie' was very interested and drawn to our skilled craftsmen, especially the sheet metal workers and carpenters, and he would often pop across to see what they were up to or show them some old tools he had acquired. On one occasion, he told us the story of his first venture into art. His wife was a very keen painter and a leading light in a Salisbury Art Society which held an annual competition. She always entered some of her works in these competitions but with a consistent lack of success. 'Reggie' decided he would have a go at painting and was quite pleased with his first effort, so he decided to enter it in the competition with the outcome that he won a prize much to his surprise and his wife's chagrin.

Lavish Entertainment

During this period, Irene and I attended a few very pleasant entertainments associated with my work. One of the officers' mess summer balls had a funfair attraction as well as the usual food and drink excesses and me aggressively driving a dodgem car with a somewhat alarmed Irene alongside in her long dress was quite stimulating.

One of the ETPS test pilot tutors was from the USA and he hosted a garden party on Independence Day, with lavish fare provided by the American Embassy.

The owner of the company who provided our design support was a flamboyant individual who originated from the East End of London, a bit like

Alan Sugar without the grumpiness. He threw very lavish entertainments for his friends, his companies' staff and customers and Irene and I attended a couple.

The first one was a hospitality marquee at the main summer Newmarket race meeting. We arrived very early so we stopped off at Newmarket town. We were looking around the market stalls when I spotted some lightweight suits and as it was a very hot day and I was feeling uncomfortably hot in the suit I was wearing, I bought one to change into. The stall holder asked us if we were going to the races and when I responded positively, he said, "Here is a tip," and handed me a scrap of paper with the word 'Dettori' written on it which meant nothing to us at the time. The hospitality marquee was very lavishly equipped with food, drink, TVs to watch the racing, a Trad Jazz band and some spectacular ice sculptures. I remembered our 'tip' and we spent the day looking for a horse called 'Dettori' while Frank Dettori won every race for which he was entered.

The following year we were invited to a party at his mansion near Stanstead. This was even more lavish than the Newmarket marquee and included a multi-course, never ending meal, which included quail's eggs, followed by entertainments which included dancing and tableaus featuring scantily clad and naked ladies. One of these was a lady in a belly dancing costume draped with a huge snake which she invited you to stroke. There were also a number of celebrities in attendance.

Music Concerts

I'm not sure now what prompted us, but in the 1990s Irene and I started attending fairly regular music concerts. The main venues we attended were the Wiltshire Music Centre at Bradford on Avon, The Turner Simms Concert Hall in Southampton, the Lighthouse Centre in Poole and various venues in Salisbury, usually during its annual music festival. The music we chose included classical, jazz and folk music.

The strangest of these was at the deconsecrated St Edmonds Church in Salisbury, which became the St Edmunds Art Centre and then the Salisbury Art Centre. The band performing was an Irish folk band called De Dannan and the concert seemed to be attended by all the Irish ex-patriots for miles around. The atmosphere was like a village hall function, including a raffle during the interval. The music was however very polished with very high levels of musicianship. Included in their set was their version of a Handel piece called 'The Arrival of the Queen of Sheba' which they called 'The arrival of the Queen of Sheba – In Galway'.

The most memorable concert was at Salisbury Cathedral during the annual festival and featured Jan Garbarek and the Hilliard Ensemble. The former was a Norwegian jazz saxophonist and the latter was a classical male vocal quartet. The music was medieval Gregorian chant type vocals with improvised saxophone above it.

Also memorable was a concert at the Wiltshire Music Centre by the veteran jazz pianist Stan Tracy and saxophonist Bobby Wellins. The music was mostly from their classic album 'Under Milk Wood' inspired by the Dylan Thomas radio play.

The most frequent performer we saw was Salisbury born jazz saxophonist Andy Sheppard, who we saw no less than four times, with various combinations, at most of the locations mentioned above.

The Rover Mini Cooper

Because Irene's volunteer jobs involved quite a bit of travelling, I bought her a Rover Mini Cooper from the Rover dealer in Andover. This proved to be a very pleasant car to drive and much better than any of my previous Mimi's. Before I did this, I organised a refresher course of driving lessons for her as she had not driven for some years.

However, her driving came to an end when we collected the Mini from a service at the Andover dealer. We drove there in convoy with Irene driving the Mini. On the return journey after the servicing, Irene lost contact with me at a road junction. She then took the wrong exit at the A303 Island and ended up heading for Basingstoke and London instead of Amesbury. She took the first exit from the dual carriageway and eventually found her way back to Newton Tony after a long anxious wait by me. This experience unnerved Irene considerably and totally put her off driving and she never drove again. The point of having the Mini gone, I decided to exchange it for a classic sports car.

The Marcos

After the abortive TVR mentioned earlier, I decided to look at a Marcos and we paid a visit to their factory at Westbury. The factory consisted of a pair of

177

WW2 Nissen huts with a Portakabin for offices. They tended to build cars one at a time to specific orders and the current model was the Marcos Mantis which was powered by the Ford Mustang engine. This was a very low, sleek and rather intimidating device and not for me.

However, they had a number of second-hand cars on offer and one of these appeared to be what I was looking for. This was WPL 815Y, a Marcos Martina. Marcos had built a number of these as body/chassis units to accept the engine/transmission/suspension from a Ford Cortina Mk 5 donor and this particular car had been professionally built by Eurosport, the Marcos agent of Sawston, near Cambridge. It had been the subject of an article in Which Kit magazine and was in 'just built' condition, so I decided to buy it.

During the nine years I owned the car I carried out a number of modifications to it. The first one was to change the carburettor, which did not have a choke, which made starting and initial running a bit hit and miss, for a standard unit, purchased from the SU Carburettor Company in Salisbury. I next decided to replace the four-speed gearbox with a five-speed unit from a Ford Sierra. I contacted a firm at Alderbury, south of Salisbury, who were willing to provide an exchange unit so I set about removing the old gearbox. The sight of me emerging from under the jacked-up Marcos dragging the gearbox, which she called 'that trumpet thing', somewhat alarmed Irene.

My final modification was to fit a 'flow-through' ventilation system. The Marcos system was appalling and simply consisted of an electric demister for both the front and rear windscreens. This resulted in very heavy condensation in the interior.

I fitted ducts from the nose air intakes to some new vents along the front windscreen and some vents under the rear windscreen into the boot, with ducts from the boot into the wheel arches. This system worked very well and solved the cars condensation problems.

Installation Manual

Having become an MoD Design Approved Organisation, there were two aspects of our design process which I found to be unsatisfactory. The first was the long-established practice at A&AEE (and RAE) of bypassing the design office for the simpler installations which utilised standard racks and pallets attached by standard methods utilising the aircrafts floor and equipment rack attachment points. The method used for this type of installation at RRE Pershore and design organisations like Cranfield and BAE was for the Design Office to issue Design Office Instructions (DOIs) instead of a full set of drawings and I would have preferred to go along this route but this was opposed by the hangar staff, supported by the Superintendent. In any case, the Mechanical Design Office was not staffed to cope with the extra workload involved, so I came up with a couple of proposals to deal with the problem.

The first of these was to train selected hangar supervisors in basic stress analysis and airworthiness clearance techniques and to this end, I organised a special course at the College of Aeronautics at Cranfield.

The second was to compile a comprehensive Aircraft Installation Manual covering all the various techniques for mounting equipment's in aircraft. I started to compile this document myself, but found that my high workload gave me virtually no time to apply to the task. I then thought of using a student engineer but these were few and far between and by the time I had explained what was needed, I could have done the job myself. The outcome of this was that work on it proceeded at a snail's pace.

Design Review

The second area of concern in our design procedures was carrying out a meaningful design review. In a conventional design review process, there is discussion between the customer, the Design Office, the manufacturing and installation staff to satisfy such criteria as value for money, quality and airworthiness/flight safety.

In a typical A&AEE instrumentation fit, the 'customer' was the flight test engineer/test pilot and his 'specification' was a list of data he needed to assess the performance of the aircraft or system being assessed. This customer had no interest in how this was achieved and therefore had no interest in being involved in a design review.

The Computer Instrumentation Group (CIG) would take this specification and compile a list of hardware, such as recorders, transducers, cameras and aircraft system tappings to provide this data. They would also provide the software needed to convert the outputs from the instrumentation hardware to the format needed by the customer.

This hardware was then handed over to my design organisation to install in the aircraft and provide the necessary wiring for system interconnects and electrical power from the aircraft power supply systems. The mechanical design team would usually produce an initial scheme and then discuss this with the workshops and hangar supervisors before producing the finalised scheme. My design team then consisted entirely of relatively young and inexperienced designers who were not that long out of their apprenticeships so they welcomed this feedback. This did not work quite so well with our design support contractor located at Stanstead. The final review was an airworthiness/flight safety assessment, including stress analysis.

I favoured a series of design reviews at each of these stages, just involving the individuals concerned, rather than a large meeting involving all concerned in the whole process. The quality organisation however favoured the latter and this was the cause of some conflict.

Breakdown

With process of Boscombe Down transferring from DGT&E to DTEO/DERA, a considerable amount of my time became involved in attending seminars on how to operate in an agency environment, including financial management, 'motivational' and human resource management material. There was a lot of 'jargon' in the documentation we were given and one of the attendees, in a state of frustration scrawled the following message on the front cover of his document, 'Also available in English'. This gave me even less time to devote to my pressing problems of the installation manual and design reviews.

I did not particularly notice the build-up of stress, apart from a lot of sleepless nights worrying about the design problems, so what followed came out of the blue. I was having a meeting with one of the quality team on the subject of design reviews when I uncharacteristically started shouting, causing him to beat a hasty retreat.

About an hour later I was called to a meeting with the superintendent and quality manager and just as the meeting got under way, I experienced acute chest pains and collapsed. I am not too aware of what happened next but it must have looked like a heart attack, so an ambulance was called and A&E at Salisbury hospital soon established that it was not a heart attack and I was sent home on sick leave for further investigation.

About a week later I was subject to a brain scan which did not reveal any problems, so I was prescribed a course of medication called beta blockers and referred to a psychiatrist. As soon as I started taking the beta blockers, I felt a strange and very unpleasant sensation so I immediately stopped taking them. I have always been averse to taking medication, even shunning aspirin/paracetamol.

I eventually saw a psychiatrist who was a pleasant individual who suggested a series of techniques for combatting stress. He also gave me some tapes of soothing music and such things as whale noises an American Indian chant, which with my tastes in music only succeeded in making me very irritable so I gave up on that approach.

After about a month of sick leave, I was getting bored and anxious to get back to work but it was about another two months before the GP allowed this.

Chapter 14
Resource Management

Resource Manager – Flight Test Services (FTS)
Resource Manager for a group of 50 engineers, scientists and technicians employed on the provision of instrumentation, data retrieval and computer services in support of aircraft assessment trials.
Resource Management Consultant
Carrying out resource management projects for the AT&E Management Cell

All change at Boscombe Down

When I returned to work from sick leave, I was in for a shock. DTEO had ceased to be a separate organisation within DERA and Boscombe Down, renamed Aircraft Test and Evaluation, had a new chief superintendent who came from the old RAE organisation. The superintendent of Technical Services had been dismissed pending an investigation of fraudulent travel claims and the organisation taken over by aircraft department, which had been transferred from Farnborough with the DRA air fleet.

The top two tiers of management at Boscombe Down now consisted of the Director of Test and Evaluation, with an engineering director heading up all the engineering services and a technical director heading up the test and evaluation activities. The resource management was carried out by a separate Management Services Cell reporting directly to the director of T&E.

The impact of all this on me was that my design and workshops organisation had been incorporated in the larger ex RAE design organisation. There was a lot of resentment among the ex TSD staff but this reorganisation was logical and if I was in the new director's position, I would have done the same.

The engineering director initially set me up as his sort of technical advisor and the first task he gave me was to look at the ex-DRA Farnborough design procedures to see if they were still suitable for the Boscombe Down situation. I found these procedures very similar, near identical in fact, to the procedures I had devised for the TSD design organisation with the only difference being the use of Design Office Instructions for simple installations. The few minor anomalies I found were discussed with the chief designer and amended as required. I therefore reported that I could see no real problems with these procedures and I think that the engineering director was a bit disappointed that I could find nothing to criticise.

I was then given the task of reviewing the work being carried out in DERA and elsewhere on unmanned air vehicles (UAVs), with a view to producing a flight trials and operational flight safety policy advice document. This work was interrupted when a key member of the chief superintendent's Management Services Cell became very seriously ill with cancer from which she subsequently died and I was asked to replace her.

The lady concerned, Kathy Strychacz, was originally recruited by me to replace the retiring engineer i/c mechanical workshops. She was a very talented and driven individual and somewhat wasted in this role and when an opportunity came in resource management, I advised her to take it and recommended her for the post. With her usual drive and enthusiasm, she soon became a key member of the Management Services Cell.

Resource Management

On being transferred to the Management Services Cell, I was asked to act as resource manager to the Computer Instrumentation Group and Avionic Test Rigs, and compile a set of procedures for carrying out this role. The Engineering Director took little interest in these units and they were left to run themselves, with the outcome that the managers of these areas looked to me for guidance on staff management and even engineering matters.

The key technical managers in the Computer Instrumentation Group were the husband-and-wife duo of Adrian and Barbara Wood with Adrian managing the mainframe computers and the provision of software for data reduction and analysis and Barbara the instrumentation hardware. There was also two small self-contained units, the Accident Data Recorder (ADR) Section and the Telemetry Section.

These were both two-man units, with the ADR Section being responsible for retrieving data from ADRs both for trials purposes and aircraft accidents. The Telemetry Section was initially somewhat under-utilised, with only high-risk trials and the annual ETPS spinning exercises utilising the facility. However, with the arrival of the DRA air fleet. The VAAC Harrier (Vectored Thrust Aircraft Advanced Flight Control) extensively used telemetry and provided the section with a steady workload.

I did not find the resource management as rewarding or interesting as engineering but I must have been doing something right because when I retired in March 2000, I was retained for two and a half years as a resource management consultant.

The Marcos, a Hawk and me on the day of my retirement, 24/3/2000

www.ingramcontent.com/pod-product-compliance
Lightning Source LLC
Chambersburg PA
CBHW051515170526
45165CB00002B/477